GETTING INTO

Engineering

SECOND EDITION

NEIL HARRIS

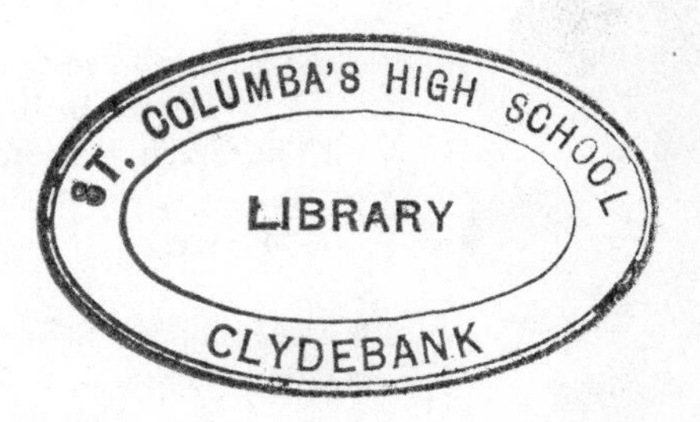

TROTMAN

Getting into Engineering
Second edition

This second edition published in 2002
by Trotman and Company Ltd
2 The Green, Richmond, Surrey TW9 1PL

British Library Cataloguing in Publication Data
A catalogue record for this book is available from the
British Library.

ISBN 0 85660 819 X

Typeset by Mac Style Ltd, Scarborough, N. Yorkshire

Printed and bound in Great Britain
by Bell & Bain Ltd

CONTENTS

About the author iv

Foreword – Professor Chris Blackhouse, Dean of Engineering, Loughborough University v

Sponsor's Introduction vii

Introduction – The fascination of engineering 1

1 The A–Z of jobs in Engineering 11

2 Getting the qualifications 26

3 Civil and Structural Engineering 34

4 Mechanical Engineering 41

5 Electronic and Electrical Engineering 47

6 Materials Engineering 53

7 Chemical Engineering 58

8 Aeronautical Engineering 63

9 Becoming a professional Engineer 67

Useful Addresses 73

ABOUT THE AUTHOR

Neil Harris is a freelance writer and career development consultant attached to the careers service at the Imperial College of Science, Technology and Medicine. Previously Director of the careers service at University College London, his other books include *Getting into the Financial Services*, *Getting into the City*, and *Getting into IT and the Internet.*

FOREWORD

The UK has always been a pioneering nation in engineering. Over half of all new patents taken out anywhere in the world are based on ideas that have originated in the UK and many global corporations look to British engineers to help them maintain a competitive edge. By reading this book you have taken the first step in what could be a very rewarding future. In discovering all the different aspects of engineering we hope that you find one that creates the spark of excitement that you've been looking for. If you decide to go on to further study you will develop a way of learning that will help you to expand your mind in all areas of life. If you enjoy problem solving you couldn't ask for a better career. Our engineering graduates are always popular with employers and the starting salary of a graduate engineer is considerably more than other graduates. The average salary for a Chartered Engineer is now more than a Chartered Accountant or Solicitor. But of course that's all in the future. For now, find out as much as you can from this book and by visiting different colleges, universities and companies. Go armed with questions about the things that are important to you and get a feel for what it is all about. If you do decide to pursue a career in engineering I'm sure you won't regret it. Good luck!

Chris Backhouse

Professor Chris Backhouse CEng
Dean of Engineering
Loughborough University

SPONSOR'S INTRODUCTION

Engineering, Technology and Science industries are major employers, contributing over £68 billion every year to the UK economy. They account for over one third of UK exports and directly employ over two million people. A further seven million people are employed in jobs which require some technology, engineering, science or mathematical skills on a day-to-day basis. That is nearly one third of the entire workforce in the UK.

The sector is at the forefront of technological development – aerospace, automotives, biotechnology, electronics, forensic science, nanotechnology and many more. The use of new technology is common in virtually every product, service or process today. The sector is thriving. It competes globally, and so the UK has had to ensure that the learning, skills and productivity of our engineering, science and technology sector improves significantly in international markets.

This has not just meant increasing the skills of people in our companies. It has meant re-skilling and multi-skilling across traditional boundaries and into new technological areas. The sector is having to work together in new ways to address key supply chain issues, becoming leaner, smarter and more productive. It is clear that the technology, engineering and science sectors face a constant challenge in updating skills to create and sustain a workforce that can continue to compete with the best in the world.

You are critical to this. Your talent, your creativity, your innovation, your enthusiasm are all needed by this vibrant and exciting sector. But where do you start? How do you see your way through the maze of options open to you? How do you get from where you are now, to where you want to be? How do you get to the top and, more importantly, stay there?

ECIS – the Engineering Careers Information Service – can help. A totally free resource for students, teachers, parents, careers advisers, companies –

in fact anyone who wants to make a difference to the world we live in – ECIS is available to provide impartial, uncluttered, sensible advice on what matters now – you. As the only resource that covers the whole spectrum of engineering, technology and science, we can see you for who you are.

We are available

- by phone (free) on 0800 282 167
- by email on ecis@emta.org.uk
- on the web at www.enginuity.org.uk
- or by post at 14 Upton Road, Watford, Hertfordshire, WD18 0JT.

You might also see us at a careers exhibition near you – feel free to come and say hello!

We are run by a charity – the Engineering and Marine Training Authority (EMTA) – so we are able to help you find the career option that is right for you and no one else. After all, we have been there, seen it and done it ourselves, so we know what you could expect.

This is why we have worked with Trotmans to bring you this book. Between us we aim to give you all you need for a great career.

Enjoy!

John, Tina and Angela
Engineering Careers Information Service

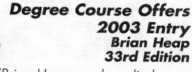

INTRODUCTION – THE FASCINATION OF ENGINEERING

As a career, engineering is grossly undersold, underestimated and misunderstood. It offers such a variety of occupations in so many different environments that few activities can match it. Jobs in engineering are so diverse that they cater for all personalities and people with many different skills profiles.

What engineers have in common is that they are involved in a process that yields tangible, usually visible results. Most engineering provides products and services that are essential to the larger community and improve our quality of life. We rely on engineering for our basic needs: water, light, heating, transport, buildings, hospitals and health treatments, food processing and so much more.

Engineering goes back a long way. The Romans were excellent road designers, enjoying quite superb sanitation and bathing facilities in addition to their underfloor central heating. In the UK our engineering traditions go back to Faraday, the inventor of electricity, Telford the bridge designer and Stephenson of 'Rocket' fame who provided the steam locomotives that gave us rail transport.

Now we enjoy the benefits of the Internet and global telecommunications, increasing use of engineered equipment in health care with a range of body scanning facilities, mechanical hearts and other implants. Engineers process and produce the medicines that are essential to keep us healthy and extend our life span. We also benefit from security systems and the means to process, store and distribute food and other goods efficiently and in first class condition. Without engineers we simply could not enjoy the quality of life that we have today, much of which we take for granted. The energy we use, whether from oil, gas, wind, water, nuclear or solar sources, is all harnessed and distributed to our homes, offices and factories for us by engineers. Engineers have a vital role to play in our society.

But it's not just the challenge and the service to society that make engineering a career worth pursuing. Engineering is a career that people can slot into at many different levels with many qualifications or few. There is a great spectrum of jobs, from assistant technician to managing director. You can start at the level that suits your current skills and knowledge and progress to positions with more responsibility that match your abilities, adding experience and additional skills along the way.

It's a career that offers many different activities in a wide variety of environments. You might be focused on the design of something new, supervising the manufacturing process, organising a construction site or visiting clients as an engineering consultant. Your persuasive skills may be to the fore if you are selling engineered products or mollifying the local community who are temporarily inconvenienced by engineering works. Alternatively, as a health and safety executive or quality manager you might be firmly insisting on certain standards being upheld. As a writer you could be writing patents, specifications, installation or maintenance manuals. In short, there is something in engineering for everyone.

LEVELS OF RESPONSIBILITY

One of the biggest attractions of engineering is that it is a profession in which there are many different levels of responsibility. Engineers can discover the level that suits their abilities and there are abundant opportunities to develop one's career by promotion from one level of responsibility to another.

It is, of course, difficult to be precise about how much responsibility any engineer has and there is a good deal of overlap. But the levels are generally described as operator, craftsperson, engineering technician, incorporated engineer and chartered engineer. In Chapter 2 we look at the qualifications necessary to progress to each level, and in Chapter 9 we discuss the professional qualifications that the Engineering Council (UK) offers to those who become engineering technicians, incorporated engineers, and chartered engineers.

GENDER

Engineering has traditionally been a male-dominated occupation but there is absolutely no reason why women should not find success in this career. Many do. There are many successful female engineers, as can be seen in the profiles that appear later in the book. Currently around 17 per cent of new undergraduates in engineering every year are female. Women are more attracted to chemical engineering, where 22 per cent of new graduates are female, than to mechanical and civil engineering where the percentages are 9 per cent and 13 per cent respectively. Just 8 per cent of all engineers now registering for the professional qualifications of chartered and incorporated engineer and engineering technician are female, but that proportion has been increasing steadily in recent years.

The Women into Science and Engineering (WISE) campaign, supported by The Engineering and Technology Board, EMTA (The National Organisation for Engineering Manufacture), and the EFF (Engineering Employers Federation), has increased the recruitment of women into engineering. Interestingly, in textiles and polymer engineering 77 per cent of the graduates are female. Annual prizes such as the Young Woman Engineer of the Year award, given by the Institution of Incorporated Engineers have served to heighten awareness that for many women engineering can be an excellent career choice.

Employers are keen to recruit female engineers and most have recruitment policies that are designed to attract them into the engineering profession.

CASE STUDY

Profile

Clare Roberts, 29 years old, design engineer Yorkshire Electricity

Clare Roberts recently won the Young Woman Engineer of the Year award presented by the Institution of Incorporated Engineers.

'Becoming an engineer has not been easy but I am determined to succeed' says Clare. 'With a strong desire to learn coupled with a confident approach to my career, I personally persuaded my Chief Executive to make me the first non-graduate to complete the Yorkshire Electricity Graduate Scheme. Since that day I have never looked back.'

Clare's first job was a clerical role but Yorkshire Electricity soon promoted her to the Technician grade. They supported her studies for the Ordinary National Certificate (ONC) and later the Higher National Certificate (HNC) in Electrical and Electronic Engineering on her way to reaching her goal of becoming a Design Engineer, all within a period of six years. Clare's current role focuses on the design of minimum-cost electrical schemes at the customer's request for high and low voltage connections.

Clare is a customer-focused engineer with a can-do attitude and innovative approach to the development of engineering solutions. To quote her employer, 'she is an inspiration to others within the industry'.

SALARIES

If one of the myths about an engineering career is that it is for men, another is that it is low paid. This is a fallacy. The Engineering Council (UK) and many of the engineering institutions undertake regular salary surveys and these reveal that engineers' pay is considerably above average. According to the Association of Graduate Recruiters pay among new graduate engineers exceeds average pay for graduates in other disciplines by around 10 per cent. Those who have gained significant and relevant work experience in engineering before completing their studies usually receive considerably more. A survey in 2001 revealed that the average earnings of engineering technicians was £30,273, incorporated engineers £35,828 and chartered engineers £45,000.

EMPLOYMENT

As an engineer your probability of being in employment is considerably higher than for other occupations. When 5.3 per cent of the workforce was unemployed in 2001 only 0.9 per cent of those registered as professional engineers were unemployed. Seventy per cent of engineering graduates go straight into employment when they leave university compared with 61 per cent of the graduate population as a whole.

The engineering sectors

One of the difficulties that people experience when they decide to opt for an engineering career is making a choice between the enormous diversity of opportunities (see Chapter 1). At the outset it is useful to think of engineering in terms of its many different disciplines – electrical/electronic, mechanical, chemical, civil, etc. The opportunities for people in each of these are described in detail in Chapters 3–8. When you decide to study engineering the best way forward is to choose one of these disciplines, although some courses allow you to start studying engineering in general and specialise later (see Chapter 2). Engineers tend to join the professional institutions that relate most closely to these disciplines.

Yet professional engineers working in industry, commerce or government tend to be employed in multi-disciplinary teams. Their job titles reflect what they do rather than their engineering discipline and these are described in the A to Z of jobs in engineering (Chapter 1). It is quite possible, for example, to be a consultant engineer, a field engineer or an engineering contractor in many different areas and engineering disciplines.

There are more than 35 institutions, each representing the interests of engineers in a specific area of activity. Details of their names and addresses are given in the appendix. Those with the largest memberships are the Institution of Electrical Engineers, the Institution of Civil Engineers and the Institution of Mechanical Engineers. They each represent thousands of members at home and abroad. Others include the Institution of Chemical Engineers, the Royal Aeronautical Society and the Institute of Metals.

The engineering institutions protect the interests of engineers in their particular discipline and provide a meeting place for people with common interests. In addition they offer a route by which trainee engineers can progress towards professional qualifications that are granted by The Engineering Council (UK). This is the body that overseas the education, training and professional development of engineers in the UK. The entire process of personal development that engineers have to go through to become professional is enshrined in a document called *Standards and Routes to Registration* (SARTOR) that is published and regularly updated by the Engineering Council (UK).

A complete list of engineering institutions, including their titles and contact details, can be found in the Useful Addresses chapter. Some represent the interests of chartered engineers, while others are for incorporated engineers and engineering technicians, the other two professional grades. Many represent the interests of more than one professional grade. Full details of what these professional qualifications mean are given in Chapter 9.

Other countries have similar organisations supervising their standards of engineering education, sometimes known as the 'formation' of engineers. In Europe these organisations have come together to form FEANI (Federation Europeenne d'Associations Nationale d' Ingenieurs) the Federation of National Engineering Associations. Through membership of FEANI a chartered engineer in the UK can receive the European professional engineering title EurIng.

The Engineering and Technology Board was formed in January 2002 to take over the wider responsibilities of promoting the engineering profession and leave The Engineering Council (UK) to concentrate on the education, training, registration and continuous professional development of engineers.

The diversity of employers

The diversity of engineering goes further than the many different engineering disciplines. The key employers also represent a vast variety of activities. You could find yourself working for a car manufacturer, a telecommunications service company or equipment producer, a petrochemical plant designer and fabricator, an oil and gas company, a transport organisation, or a manufacturer of consumer goods – to name but a few! Employers in these sectors are often members of a trade association such as the Society of British Aerospace Companies and the Association of Engineering Consultants and Water UK. Often it is possible to discover the names of employers with interests in a particular field by looking at the Website of their trade association.

AN INTERNATIONAL PROFESSION

Another major attraction is that engineering is an international profession. Leading firms including Ford, Siemens, Shell, Rolls-Royce, Airbus, Nokia and Ove Arup all operate on an international basis. They have production plants in many countries and marketing, maintenance and repair operations in many more. Engineering consultants and contractors work on an international scale taking on projects world-wide. Current examples include BAe Systems, providing air traffic control equipment for Tanzania and Rolls-Royce having a contract to maintain the aircraft for Brazil's leading airline. The engineering consultant Ove Arup has offices in eight European countries and 19 outside Europe employing 6,000 staff of whom half are engineers.

SKILLS

If you think that engineering could be the career for you, consider whether you have or could develop the skills that engineers need to do their jobs successfully. Whatever their area of work all engineers have to be interested in the application of science. An interest, and qualifications in science, even at a fundamental level, will help you to understand engineering. Electrical and electronic engineering is based on physics while chemical engineers need a thorough understanding of chemistry. All engineers need to have some mathematical ability, though structural and aeronautical engineers are required to have more than most.

Engineers are practical people. They seek and find practical solutions to problems that usually involve the manufacture of engineered parts resulting in tangible products. By no means every engineer works on the making and fitting of pieces of equipment or their parts, but most have to understand how it's done.

There is a curious paradox about engineering. Some aspects of it are very conservative, rooted in tradition and experience. Mature professionals know what works and what doesn't, what is safe, tried and tested, and. they tend to stick to it. Yet engineering is also about creativity and ingenuity, designing and making new things that work and meet a particular need. You don't have to look far to see engineers at the cutting edge of technology. They can be employed in the design of medical applications, satellites, optical telecommunication systems using lasers, or new wind farms. Whatever the application, engineering is a continual challenge. It is a career in which you have to keep abreast of and adapt quickly to fast-changing technologies. Adaptability and flexibility are among the skills required of most engineers.

Communication skills are also vital. Engineers communicate by sketches and drawings, plans, reports and simply talking about the problems they face. But they have to be able to get their ideas across to technical and non-technical people alike. A good engineer will use the industrial jargon when talking to colleagues and explain matters in simpler terms when addressing clients or managers from non-engineering areas.

NATIONAL VOCATIONAL QUALIFICATIONS

Previously, every industry had a body, known as a national training organisation (NTO) that was responsible for the skills and training of its workforce. EMTA (The National Organisation for Engineering Manufacture) occupied that role for the Engineering industry and offered workplace qualifications that cover the aerospace, electrical/electronic, mechanical engineering and metal trades as well as the motor manufacturing, ship building and marine engineering organisations.

It is currently (2002) being transformed into one of the new Sector Skills Councils which are replacing NTOs. Its mission is 'to raise the level of skills in the British engineering industry to world class standards'. This is achieved by setting detailed standards for National Vocational Qualifications from the craft level (NVQ 1) to managerial competencies (level 5). It is responsible for maintaining the standards of these qualifications and encouraging employees to develop their skills through attaining them. Chapter 2 gives more details of how their work impinges on the training of engineers.

SCHEMES FOR YOUNG ENGINEERS

There are numerous schemes designed to help young people who wish to get a flavour of the excitement of a career in engineering. The Engineering and Technology Board, www.etechb.co.uk, provides a monthly electronic newsletter 'efirst' on its Website that includes all the latest news and information. The Young Engineers for Britain competition provides opportunities for those under 18 to collaborate with a local engineering company and pursue a project and enter it into this national competition with regional heats. Last year's winners, Brendan Quinn and Enda Young worked with Northern Ireland

Electricity to solve the problem of birds settling on power lines and leaving their droppings on the places beneath. They devised a self-sustained induction defence system, a device that uses the power from the cables to move quickly up and down them and scare off the birds.

A similar scheme for under-11s is known as the Junior Engineers for Britain competition.

The Year in Industry scheme helps students who want to gain a year's industrial experience between leaving school or college and going to university. Many employers in the engineering industry support the scheme and participation in it can lead to sponsorship throughout a degree course.

European Young Engineers provides opportunities for student engineers to have contacts with similar students in other countries, including attending conferences. The organisation boasts a membership of 100,000 young engineers, mostly university students in engineering, in several European countries including Belgium, Denmark, Finland, France, Germany, Ireland, the Netherlands, Hungary and the UK.

The International Association for the Exchange of Students for Technical Experience is an international organisation that provides employment opportunities for student engineers and scientists to work in countries other than their own. It is an exchange scheme, so when a student from one country takes a temporary job in another, a second student goes in the opposite direction.

HOW TO USE THIS BOOK

This book is designed to give you a thorough insight into what it means to be an engineer. It includes details of how to get started on an engineering career, the qualifications you need and the different routes your career could take you. Our A–Z of engineering jobs (Chapter 1) demonstrates the breadth of opportunity and the fact that there is work to do in engineering to suit a wide variety of interests, personalities, and ranges of skill. It lists the job titles many engineers have and gives details of what they do.

We also take a more in-depth look at all the key engineering disciplines in Chapters 3–8 – aeronautical, civil, electrical/electronic, mechanical, materials and chemical, discussing in which sectors each type of engineer primarily finds opportunities and the major employers in each sector. At every stage we give details of the qualities employers seek in their recruits.

Finally we take a close look at what it means to be a professional engineer, the different professional qualifications and how people study and train to gain them. If you want to know about one particular aspect you can go straight to the relevant chapter. If your knowledge of engineering is basic you might start at the beginning and work your way through the book to find out which areas interest you most. Whatever your starting point we hope this book conveys the challenges, the fulfilment and the job satisfaction that a career in engineering can provide.

Chapter 1
THE A–Z OF JOBS IN ENGINEERING

A fully comprehensive list of engineering jobs is not possible within a short book such as this. Below, however, we have given details of the main engineering job titles. These are occupational roles and should not be confused with the engineering disciplines such as civil, mechanical and electronic engineering described in Chapters 3–8. Building services, for example include electrical wiring for lights, phones and computers that may be the province of electrical engineers but also air conditioning, ventilation, lifts and escalators that are usually dealt with by mechanical engineers. Engineers from any disciplines can become consultants, designers or sales engineers, or provide customer services.

AGRICULTURAL ENGINEER

Agricultural engineers are concerned with machinery for the farm. They research the needs of farmers and design, develop and produce equipment to meet those needs. This includes the design and manufacture of harvesting equipment, tractors, milking machinery, etc.

Some agricultural engineers are responsible for the maintenance of machinery once in operation. Contact the Institution of Agricultural Engineers for more information (see Useful Addresses).

BUILDING SERVICES ENGINEER

Building services include the supply of water, electricity, gas, heating, ventilation, lifts, lighting and the networks for computer equipment. Usually they are so well hidden in the areas between floors that most people never see them, yet no building could function without them. Building services have to be designed into a building from the outset and planned before construction. Inevitably, new services have to be introduced during the life of a building and must be designed and installed. There are many jobs in the maintenance and repair of building services.

COMMISSIONING ENGINEER

When a new piece of equipment or processing plant is being installed the people in charge of this activity are often called commissioning engineers. The term is often used in the petrochemical industry where new processing plant is built and commissioning engineers are responsible for getting it to operate. Their task is to optimise the operation of the entire plant so that it produces product to the required specification at minimum cost and with as little waste as possible. Commissioning engineers sometimes work in manufacturing, where a new piece of equipment is installed that must be integrated with what is already in service.

CONSULTING ENGINEER

Consulting engineers are employed by clients who need advice from fully qualified engineers. While many consulting engineers work in the area of construction and civil engineering, there are also many in other sectors, especially mechanical, electrical and control engineering, building services and safety. The work is project-centred and often includes many different functions such as design, costing, logistics and the supervision of construction. The large consulting engineering firms work on a global basis.

CRAFTSPERSON

This is a role in which an engineer installs electrical circuits or produces parts of machinery and fits them into the whole assembly (see Fitter). They interpret engineering drawings and use lathes and other equipment to turn those designs into products.

CUSTOMER SERVICE ENGINEER

When people spend large amounts of money on engineered equipment they expect to get a prompt and efficient after-sales service should anything go wrong. These engineers provide an after-sales service to customers. This can take many forms, from providing a workshop where equipment can be brought for repair, to visiting customer sites to maintain their firm's products. Some customer service engineers, especially in the software and computing sector, provide telephone help lines for their customers. In these situations there are often three layers of engineer: first, the one who deals with problems that can readily be solved; next an engineer who takes on problems which will take some time; and third, those who tackle the really difficult problems that arise.

DESIGN ENGINEER

Design engineers are often seen to be at the beginning of an engineering process, designing the new model, the latest construction. In fact they have a responsibility for a product throughout its lifecycle and must be prepared to make modifications to their designs as time goes by. These engineers must choose the materials, have ideas about how the product would be made and concern about whether customers will want to buy the product. They need to be innovative and creative with an excellent knowledge of their sector of engineering. They must also liaise well with researchers, developers, production managers and the marketing department. Computers are usually employed extensively in the design process and there are several design software packages such as AutoCAD.

DEVELOPMENT ENGINEER

Any idea for a new product has to be developed before it can go into the production phase. Development engineers are concerned with the equipment required to make a product, the cost of raw materials and compliance with legislation. Can it be made economically? Will it be safe? Do we have the equipment to make it or should we get someone else to produce the parts? Will we need to modify our existing production plant and how? These are the questions with which development engineers have to grapple.

DRAFTSMAN

Drafting involves the preparation of plans, engineering drawings and specifications. Some of these activities are supported by computer systems that offer the opportunity to model things in two or three dimensions and then draw different perspectives.

DRILLING ENGINEER

There are many different circumstances in which it is necessary to drill into the ground. Oil wells and artesian wells must be drilled. Samples of soil are usually required when assessing the ground on which foundations are to be laid. Oil companies employ specialist firms of drilling engineers, who may be drilling a hole 7,000 feet deep and under 300 feet of water. The engineer in charge of the operation is a drilling engineer and their assistant is sometimes called a 'roughneck'. Drilling engineers are responsible for all the equipment they require and for keeping it maintained. They also have a detailed knowledge of the different techniques that can be used in the many situations they face. Oil wells, for example, are not drilled in straight lines and have to travel through many different geological strata, which makes this a highly skilled job. Engineers may encounter pockets of gas and high pressure, which also pose problems and have their dangers.

ENGINEERING SURVEYOR

There are many instances in which an engineering work must be surveyed, and this is the domain of the engineering surveyor. One key area is that of insurance. When a ship, aeroplane or engineering facility has been insured and something goes wrong, leading to an insurance claim, an engineering surveyor is usually sent in to assess the damage. Surveyors often work with loss adjusters who act as independent assessors of insurance claims. Another source of work is the surveying of ground before construction work begins. This may include analysis of the rock or soil – known as 'geotechnical engineering'.

ENVIRONMENTAL ENGINEER

Environmental engineers work on the design and construction of systems that distribute water and collect waste material, including sewage. Their initial training is usually in civil engineering. The term is sometimes used to describe those who perform environmental impact tests, investigate contaminated land and related environmental issues such as air and water quality.

FIELD ENGINEER

A field engineer is simply someone who is out and about. Their job is usually concerned with installation, after-sales service, maintenance and repair – see Customer Service Engineer. Field engineer is also the term given to someone who provides an engineering service away from their base, such as seismic surveys.

FITTER

A fitter is an engineer who assembles parts to make an entire article. Each part must be made to a prescribed specification and fit into the whole piece exactly in the way it was designed to do. If not, the fitter may have to make adjustments, perhaps by machining, drilling or the use of other shaping tools. A fitter is often an engineering technician.

HEATING AND VENTILATION ENGINEER

These engineers are focused on building services, their installation, maintenance, adaptation and updating. See Building Services Engineer.

INFORMATION ENGINEER

A relatively new role, these engineers specialise in collecting, analysing, processing and distributing information. Using such techniques as relational databases, data warehousing and mining, they organise data so that they can be usefully explored and assist in decision making. Modern databases can be extremely large, such as all the customers of a bank or details of the human genome. It is not humanly possible to search them for information without the techniques developed by information engineers. They are sometimes called information systems engineers because they organise systems that allow such investigations to be made.

JOINTING ENGINEER

A jointing engineer is responsible for joining different pieces of a structure together using welding, riveting, bolting and similar techniques. Different methods are required in different situations, especially when joints are subject to dynamic or thermal stress.

LIGHTING ENGINEER

These engineers have a thorough understanding of illumination and how to provide adequate lighting in many differing circumstances. Their work covers street lighting and floodlighting as well as providing such services for offices and the home.

MAINTENANCE ENGINEER

Regular and planned maintenance avoids critical, perhaps unsafe situations arising. In the production process, maintenance, often carried out during unsocial hours when plant is closed, ensures continuous manufacturing during normal production hours. Maintenance engineers ensure the optimal running of the machinery for which they are responsible by planning ahead and obviating the need for repair. They have an expert, intimate knowledge of the machinery they maintain and a thorough training in all aspects of its operation.

MEASUREMENT ENGINEER

Accurate measurement is often essential. When we buy products such as petrol we expect the pumps to measure what we get as exactly as they can. Weights and measures officers quickly deal with those who sell short measures. But measurement is also important in other fields including the calibration of measuring equipment in health care situations.

MECHANIC

Mechanics maintain and repair mechanical equipment, often by dismantling it, replacing faulty parts and assembling it again into full working order. In addition to a basic education in engineering, many mechanics are trained specifically to work on certain products, often through attending training programmes provided by manufacturers. When mechanics are professionally qualified they are usually called engineering technicians

MUD LOGGING ENGINEER

When an oil well is being drilled the drill is cooled with lubricants known as 'mud'. Analysis of the contents of the mud provides information about the geology of the rock through which the well is being drilled and especially whether or not it is oil-bearing. People working in this area tend to have geological qualifications, but see Wireline Engineer below.

NETWORK ENGINEER

The growth of telecommunications over the last decade has been exceptional, leading to a dramatic increase in work for network engineers. There are several types of network. First are those of the telecommunications services providers such as BT and NTL. Then there are private networks such as are found in a bank or a large organisation. Local area networks (LANs) cover a small office while wide area networks (WANs) provide communications between offices or factories in different locations. Firms that provide and install equipment for networks employ network engineers. Some also work for telecommunications service providers and organisations that have their own networks.

OPERATIONS ENGINEER

An operations engineer is someone who is in charge of, or working on an operation. This can mean many different things. It can describe the person responsible for getting a North Sea oil installation from a harbour to its right location and then tipping it up into a vertical position. It may, however, be an engineer in charge of a section of rail track, the mining of a quarry or the laying of a pipeline. There is usually an engineering task to be completed at a specific location and the operations engineer sees to it that it is completed.

OPERATOR

This is the term for someone who operates a piece of equipment. They may supervise a part of the manufacturing process or take responsibility for some workshop tools that perform a specific range of tasks.

OVERHEAD LINE ENGINEER

Electrical power has to be distributed from power stations to our homes, schools, hospitals, offices and factories. At the outset it travels down overhead cables and later it reaches us by underground transmission. People who install, maintain and repair overhead cables are called overhead line engineers. These engineers also work on the electric railways. If this work interests you, a good head for heights, willingness to work outdoors in all weathers and at night and a fundamental knowledge of electrical engineering are the key requirements.

PETROLEUM ENGINEER

These engineers are responsible for getting the maximum return out of an oil field in terms of oil extracted. They analyse geophysical data to decide where wells should be drilled and what methods of extraction should be used. They calculate how much oil and gas will be yielded and whether the field is economic, and they also supervise the entire operation of oil well drilling and production.

PLANT ENGINEER

A plant engineer is usually responsible for a processing plant, in contrast to a manufacturing production facility. Where chemicals react and distil or fluids move continuously through pipe-works the operation is often managed by chemical engineers. Water, margarine, cleaning fluids, soap and perfumes are among the products that are manufactured in this way. The plant engineer is responsible for optimising the output of the plant in terms of quality, energy employed and waste. Treating effluent so that it is environmentally acceptable and can be removed safely is often a part of this role. Since plant is usually automated the engineer has to be conversant with electronic control systems. Senior plant engineers supervise the entire operation, including staffing, the throughput of materials, packing and distribution of the product.

PRODUCTION ENGINEER

Manufacturing requires a whole range of engineering. It usually includes belts that move things along. There may be containers that are cleaned, filled, capped and labelled. Cars are welded and sprayed with paint. Electronic circuits have their parts installed by robot and systems that solder items into place. Whatever the article being made there is a lot of engineering involved. Production engineers ensure that the machinery required for production is working and provides the maximum output at the right quality and with minimum waste.

Firms that make production equipment also employ these engineers. They design, develop, make and install it at their clients' premises. Those working on the production line itself decide what machinery they require to make a product, often in consultation with development and design engineers. They work in teams with other engineers and may have to work when the production line is not in operation (weekends and nights). Each piece of equipment must operate in harmony with those preceding and following it so that the flow of product is not interrupted and no bottlenecks occur.

PROJECT ENGINEER

Much of what goes on in engineering is achieved in projects. Whether it is the construction of a bridge or the building of a petrochemical plant, the work is often undertaken as a project that will continue until completion. A team of engineers is brought together for this purpose, each with their own contribution to make, and disbanded when the project is finished. The person in charge is often called a project engineer.

The job is often very broad. It may include planning the project and working out the logical sequence of events. Project engineers are sometimes responsible for negotiating with suppliers of equipment and subcontractors who will complete a part of the work. They also have to communicate with the client to be sure that their wishes are met and monitor progress to keep the project on time and within budget. In short, this is a wide-ranging job that needs experienced engineers who have a specific knowledge of the engineering involved but who can also handle all the other aspects of managing the progression of a project.

QUALITY ASSURANCE ENGINEER

The production of quality products is essential for the survival of engineering companies. There are two aspects to this. The first is to plan quality into the design and check that what is produced achieves the correct specification. Quality assurance engineers assess these outcomes and take action to rectify the situation when things go wrong. But many production plants have an attitude of total quality instilled into their staff. This means aiming for continual improvement, to reduce waste, increase efficiency and produce more goods of the right quality first time, so that fewer have to be reworked in order to bring them up to the required standard. Quality engineers are involved in all of these situations.

REFRIGERATION ENGINEER

Refrigeration engineers are employed all along the supply chain in the food industry to keep foods at the optimum temperatures. This includes being involved with refrigerated vehicles, warehouses, shops and sometimes ships. Breakdowns in refrigeration can lead to the loss of huge quantities of stock, making this an essential role. Techniques used in this area of work are also employed in air conditioning systems.

RESEARCH ENGINEER

Highly qualified engineers and technicians are often attracted to work in research. Some are employed in universities and some in contract research organisations. Many of these are members of the Association of Independent Research and Technology Organisations (AIRTO) and each serves different industries. Examples include RAPRA, serving the rubber and plastics industry, EA, investigating problems for the electricity industry, and SIRA, offering a similar role in scientific instruments. Many large companies, such as Shell and BT also maintain their own research departments. In industry the job is to investigate novel ideas that may lead to improved efficiency, new products or processes and be self-financing within a short period. In universities the work can be more academic and investigations do not necessarily have to lead to immediate financial gain.

SAFETY ENGINEER

Safety is fundamental to everything that is done in engineering and of paramount importance. When things go badly wrong in this area the entire business is at risk, as has been demonstrated most vividly by the recent problems at Railtrack.

In every engineering organisation someone is responsible for safety and in larger ones there is usually a safety group or a committee that oversees all safety issues. When things do go wrong the Health and Safety Executive, which also employs engineers, is called in to decide why and take appropriate action, including closure of the plant.

Safety engineers have to be conversant with safety regulations and ensure that all staff obey them. Among their functions is the completion of safety audits on current operations and having an input to proposals for forthcoming developments. Training new staff in safety matters and maintaining safety awareness among all personnel are also a part of their remit.

SALES ENGINEER

When engineered products are sold, especially when they are bought by businesses, the sales person is often an engineer. The sale is usually negotiated with engineers and the sales person must have a detailed understanding of the engineering that underlies the operation of the product. Additionally, a good sales person has the ability to understand their customer's needs and this is often easier if they have a good level of engineering education and experience.

Excellent interpersonal skills, an ability to communicate effectively and negotiate persuasively are, of course, essential to this role and, since the job usually involves visiting customers at their premises, a driving licence is often required.

SITE ENGINEER

These engineers are responsible for a site, or part of a site where there is engineering activity. In addition to being concerned that what happens is in line with the engineering drawings and specification, they may also be responsible for managing the staff, organising the schedule of work that is carried out, taking care of security, supervising subcontractors and many other functions. The job of site engineer is multi-faceted and involves the exercise of considerable management skill in addition to engineering expertise.

SOFTWARE ENGINEER

Software engineers are concerned with programming of software that is embedded in equipment such as telephone systems. They build large suites of programs from basic building blocks of software that have been tried and tested, making it easier to update, maintain or alter. Often these are available in software libraries and form the kernel of a new system. When working with electronic engineers they decide which functions of a circuit will be carried out electronically and which via the software. Each prepares their own part of the design and then puts the two together for testing and eventual implementation.

SOUND ENGINEER

Sound engineers work in broadcasting and recording studios to optimise the acoustics and take responsibility for the electronic systems that detect, analyse and filter sounds. Some are employed by companies that produce equipment for these purposes. An education in electronic engineering or physics is usually required. The Institute of Acoustics is the professional body of sound engineers.

SYSTEMS ENGINEER

Products such as cars and aircraft include many different systems, all of which interact with one another. These include hydraulics, avionics and electronics, power (known as powertrain in cars, meaning the method of transferring energy from the engine to the vehicle), lighting, etc. Systems engineer is the term often used for people who have responsibility for a particular system. Their job is not only to ensure that the design, development, operation and maintenance of the system are achieved cost-effectively and smoothly but also to integrate it into the overall system for the product.

TECHNICAL AUTHOR

Technical authors write essential documents for engineered products. These include specifications, installation and training manuals and maintenance instructions. The job requires an ability to write lucidly, succinctly and unambiguously about an engineered product. It is essential for these writers to have a thorough understanding of the product and how it will be used. Technical writers usually have a degree in engineering.

TECHNICIAN

Many engineers are described as 'technicians', a role that includes a range of activities. It may mean being responsible for a specific service, eg audio-visual technician, and having a detailed understanding of the engineering involved. In other circumstances the role is to be an assistant to a senior engineer. The level of responsibility in these roles varies considerably. Some technicians are incorporated engineers and others engineering technicians.

TENDERING ENGINEER

Engineering organisations, especially consultancies, have to seek work continually to survive. If a chemical company, for example, decides to construct a new plant or a football club wants new floodlights, they will invite tenders for the work. Engineering organisations develop strong relationships with their clients that provide them with opportunities to tender. Then they develop designs and specifications, calculate the time required and the costs involved. These result in a tender document and often a presentation by a team of engineers during which they demonstrate that they have the capability to undertake the work. If the costs of the proposals are too high they are not accepted, and if too low they could end in bankruptcy for the firm, so tendering engineers walk a thin line between the two.

The job includes talking to potential clients, gathering large amounts of diverse information, and discussing various aspects of the problem with colleagues before writing and presenting the proposals.

TEST ENGINEER

This role consists of testing engineered products to make sure that they work satisfactorily. This may be at the end of a production line or relate to products that have been sent for maintenance. Some testing procedures, for example testing a computer or control system, are highly sophisticated and extensive, either because the product is complicated or for safety reasons. Test engineers are sometimes responsible for developing new test and installation equipment and training others in their use.

WIRELINE ENGINEER

Wireline engineers dangle equipment down oil wells that will carry out physical tests on the surrounding rock. They might measure the density, porosity, resistivity and many other properties that give clear indications about the formation of the rock and in particular whether it is oil-bearing. Equipment can also be used to measure the pressures and temperatures at the bottom of oil wells and measure the flow of oil, water and gas. See also Mud Engineer.

WORKSHOP MANAGER

Items often have to be taken to a workshop for repair. Workshops are set up for many different reasons including the use of lathes, drills, welding equipment and related services. Workshops are also set up to modify parts when fitters experience problems with their assembly. The workshop manager is responsible for the upkeep and maintenance of the tools in the workshop, the flow of work through it and the supervision and training of engineers working there.

Chapter 2
GETTING THE QUALIFICATIONS

So you want to get into engineering. What qualifications do you need? There are numerous qualifications and they will all help you to get into and develop a career in engineering. You can begin with basic qualifications and build on them later depending on the level of professionalism and seniority you want to reach. It's a system of ladders and bridges that can take you as far as your abilities will allow.

Let's start at the beginning. A good understanding of basic maths and physics is essential for most jobs. Engineers do lots of measuring, calibrating and calculating. They produce engineering drawings and plans. They use geometry on a regular basis and some get involved with models for complicated matters like fluid flow and thermodynamics. Engineers have to choose the best material for the job and that entails understanding the physical properties of what you are using.

LEVELS OF QUALIFICATIONS

Operators

If your goal is to be an operator of a machine or equipment a GCSE in maths, science, applied science, engineering, manufacturing or ICT is a useful qualification. Another route is to take a GNVQ at foundation level in engineering, manufacturing, science or ICT. A BTEC First qualification is also recognised by employers. At work your training will usually include the completion of a National Vocational Qualification to level 1 and level 2.

Craftspeople

Aiming to become a craftsperson you can start by gaining several of the relevant GCSEs mentioned in the previous paragraph. An alternative route is to study for the Intermediate GNVQs or a BTEC First. Armed with these qualifications you can apply for a Modern Apprenticeship and get valuable training at work that will lead to a National Vocational Qualification at levels 2 and 3. The City & Guilds qualifications are craft based and help to develop specific craft skills that are required in the workplace.

Most of the leading engineering companies have an excellent attitude to training. If you get a job straight after leaving school many will send you to college on a part-time basis (see Westland in Chapter 7). There you might study for a BTEC First or Ordinary National Certificate in engineering.

Engineering technicians

Engineering technicians may commence their education in the same way as craftspeople and operators but eventually they will need to gain an AS- and A-level or a Vocational Certificate of Education (VCE; previously known as advanced GNVQs). Again the subjects studied might be engineering, electronics, construction and the built environment, design and technology, ICT, manufacturing, sciences or mathematics.

Should you choose the vocational qualifications route into a career or the more academic? That's not always an easy choice. Generally speaking vocational qualifications help you learn how to do things while academic ones give you an understanding of why things happen. The first is more practical and the second more theoretical, but they overlap in content. Choose the route that best suits your approach to solving problems.

Incorporated engineers

To reach this professional grade the completion of three years of study for a Bachelor of Engineering (BEng) degree or its equivalent is the ultimate academic goal. Preferably, though it is not essential, the degree course will be accredited for this purpose by a relevant engineering institution such as the Institution of Incorporated Engineers. Entry to these degree courses is gained via A-levels or VCEs or equivalent qualifications in relevant subjects.

Another route is to take an HND in the subject you wish to study for a degree. This is a two-year vocational course that can be studied at most of the new universities and Colleges of Higher Education. There is a very wide range of subjects to choose from and they are all listed in the *UCAS/ Trotman Complete Guide to Engineering Courses* published by Trotman.

Foundation and Access courses provide another route into an engineering degree or HND. Students who take A-levels in the wrong subjects for engineering, such as English, history and French yet get good grades, sometimes enrol on 'foundation courses'. These are one-year courses that cover maths and science up to A-level standard and provide a springboard into a university engineering course.

Access courses are specifically designed for mature students and those whose social background makes it difficult for them to gain entry to university by the well-trodden routes that many 18-year-olds take. These courses aim to bring participants up to a good A-level standard in maths and the sciences so that they can successfully tackle these subjects in an engineering degree or HND course and subsequently make a career as an engineer.

The Higher National Certificate (HNC) is a course at the same level as HND that is usually taken on a part-time basis, often by engineering technicians wishing to become incorporated engineers. However, in Scotland the HND is a higher academic qualification than the HNC, taking two years compared to just one.

If you successfully gain a place in the top half of the class it is usually possible to transfer to the second year of the degree course on completion of the HND. If the BEng is not accredited by an engineering institution, further study may be necessary in order to qualify as an incorporated engineer.

Chartered engineers

Chartered engineers must gain a qualification equivalent to a Masters in Engineering (MEng) after four years of university study. To gain a place on these courses good A-level grades, or their equivalent, are essential. Aspiring chemical engineers need an A-level in chemistry. If maths is your best subject consider a degree course in aeronautical or structural engineering, both of which use maths more extensively than the other engineering disciplines.

Before choosing your course it is wise to look up the relevant engineering institution on its Website and discover which courses are accredited. All the institutions regularly review engineering courses that are relevant to their discipline and give their official stamp of approval to the best. For a full list see the Engineering Council (UK) website: www.engc.org.uk.

DEGREE COURSES IN ENGINEERING

Degree courses in engineering vary a great deal (see the *UCAS/Trotman Complete Guide to Engineering Courses* published by Trotman). Some lead to a Bachelor in Engineering (BEng) degree after three years of study and some to a Master of Engineering (MEng) after four. These times are extended if a sandwich course including industrial training and experience is taken.

The MEng adds both a greater depth of knowledge in the engineering subject and more breadth in terms of the subjects studied. These don't just include engineering subjects but also prepare students for management by offering modules in areas such as finance, marketing, project management or human resources.

Often it is possible to start out on an MEng course and transfer to a BEng and vice versa, usually after the second year of study. In general it is those with the highest pre-university qualifications who are admitted to the MEng courses and those who are in the top half of their BEng class after two years of examinations who can transfer later.

Sandwich courses

From an employer's point of view an ideal candidate is one who has academic knowledge and industrial experience combined. Sandwich courses are designed to provide both, though many students on other courses get their experience by finding relevant employment during their summer vacations.

There are several kinds of sandwich course. The most popular are those which provide one year of industrial experience after two years of study at university. Students get paid work during their industrial placement and then return to their studies for a final year before commencing their careers. Quite popular among these are those that provide experience in another country and also offer language training. Increasing numbers of international employers recruit engineers across Europe and the ability to speak in more than one European language can be an important asset. For membership of FEANI (see the Introduction and Chapter 9) it is essential.

A large number of universities offer sandwich courses that provide a year of experience overseas. Most have close links with firms and universities in other European countries but some also offer the opportunity to visit other parts of the world such as Canada and the USA.

Two other varieties of sandwich courses are the 'thick' sandwich and 'thin' sandwich. The 'thick' variety involves a year with an employer before going to university and a second one once your academic studies are complete. Usually these are tied to sponsorship: the employer agrees to sponsor the student after they have got to know them for a year and feel comfortable with the relationship. The 'thick' sandwich often includes work experience in the summer vacations during the course. Some feel that this is a good way to get work experience and it avoids the effort of seeking a new job every summer. Others prefer to be free to discover a set of different employers each year and gain a broader range of experiences.

The 'thin' sandwich is a system used by Brunel University, offering alternate periods of six months with an employer and at the university. Its supporters say that it allows academic study and the application of knowledge to real industrial situations to go forward together. Its detractors feel that the continual change of scene is unsettling and doesn't enable students to focus for long enough on any one aspect of their studies or training.

Degree subjects

There is a wealth of degree subjects that can be studied. The traditional engineering subjects are aeronautical, civil, chemical, electrical/electronic, production, mechanical and materials engineering. But many universities offer specialist degrees such as air transport engineering and aerospace materials. In electronics there are degrees in medical electronics, microelectronics (integrated circuits built on a chip), electronic materials and aeronautical electronics, telecommunications engineering, digital systems and control engineering.

Students interested in chemical engineering may also find a challenging career by studying biochemical engineering or biotechnology. The engineering of micro-organisms and their processing to produce beneficial products is an important growth area. Some universities include these studies within chemical engineering courses and others offer special courses focused on the subject.

Aspiring civil engineers can decide to study structural engineering instead and still make a career in the construction industry. It is possible, however, to graduate in civil engineering and convert later into a structural engineer.

Another possibility is to study a two-subject degree, known as joint honours, which includes engineering and another subject such as management or a foreign language. A degree titled 'Mechanical Engineering *and* Management' means that mechanical engineering and management are both studied to the same depth while a 'Mechanical Engineering *with* Management' course mostly consists of engineering with a few courses in management.

Care should be exercised when choosing such courses because too much dilution of the engineering content will mean they are not accredited by the engineering institutions. If subsequently you decide that you want to register as a professional engineer you may discover that further study is necessary. However, please note that not all jobs in engineering absolutely require chartered or incorporated engineer status.

At some universities it is possible to study for a general engineering degree, at least to begin with, and then to choose the area in which you will want to specialise later. This is a particularly valuable strategy if you are undecided at the outset as to which area of engineering most interests you.

It is essential always to read the prospectus in detail of any course you contemplate studying.

Getting a place at university

It is quite normal for applicants for places on university courses to be invited to interview. If you are attending such an interview always read the course prospectus in full beforehand and go prepared to ask questions about what you would study. Use the interview to clarify the options that will arise during the course to pursue a range of different interests. Ask about the department's industrial contacts.

Show your strong motivation by being ready to talk about the engineering you have seen. If applying for a course in civil engineering you may well be asked what examples of civil engineering you saw on the way to the university. When seeking a place on an electronic engineering course be ready to discuss the latest trends that you have noticed in that area such as new mobile phones, the latest video recording equipment and related issues.

Postgraduate courses

One way of gaining specialist expertise is to study for a diploma or postgraduate Master's degree after graduating with a BEng or MEng. A Master's degree takes one calendar year full time and a diploma just the nine months of an academic year. Many can be studied on a part-time basis over a longer period. After a degree in civil engineering, for example, it is possible to take a course in concrete structures, structural engineering, irrigation, water resources, foundation engineering and many more relevant subjects. Electronic engineers can specialise in telecommunications, digital signal processing, microprocessors, networks and so on.

There are also interdisciplinary courses that engineers often move into. Robotics, manufacturing, quality and reliability, industrial design and energy engineering are just a few examples of these.

PhDs

PhD studies are for those engineers who want to make their career in research or academia. They investigate the leading edges of technology and take it forward through innovation and creative expertise to yet further horizons.

Doctoral studies are often concerned these days with real industrial problems and are conducted in liaison with engineering companies. They are funded by research councils, industrial companies and sometimes charities.

Chapter 3
CIVIL AND STRUCTURAL ENGINEERING

CIVIL ENGINEERS

Civil engineers work on all kinds of construction, from roads and motorways to railways and airports. Bridges and tunnels, dams, docks, water distribution and sewage systems, shopping centres, airports and office blocks are all within their remit.

Studies

During their studies they focus on subjects such as surveying, ground engineering and soil mechanics, all necessary for setting out the levels on a site and deciding what foundations any construction must have to be stable and firmly based. Mechanics and structures, materials and water engineering are also among their core subjects. Civil engineers have to know how to deal with water and drainage and some specialise in water collection and distribution as well as the treatment plant for effluent and waste. They also need a clear understanding of structures and how to design ones that are safe in each particular circumstance. Degree course studies usually end with a design project and the choice between specialisms such as transport, water engineering and structural design.

Contractors such as Laing, Kier and Mowlem often take school-leavers as trainees and encourage them to progress towards graduate qualifications. Some provide sponsorship to this end but also seek to recruit new graduates in civil engineering. Most engineers working in this area have at least a higher national diploma.

Main areas of work

Consultants

Consultants discuss their clients' needs and translate them into designs and costings. They liaise with the contractors who win the opportunity to undertake the construction work and supervise what goes on at the site. Most people employed by consultants are graduates, often with postgraduate qualifications, and they usually begin their careers by working on detailed designs for specific projects. When they have become skilled in that area they begin to supervise parts of a project that they have been responsible for. Geotechnics, for example is the science of earth. Whenever something is to be constructed on a site the area is surveyed, its levels mapped and the earth analysed so that engineers are sure of the foundations they will build.

Eventually their expertise is applied when they visit construction sites and check that what has been built is in line with the specification and the drawings and within the set budget. Senior engineers concern themselves more with developing relations with clients, producing work that meets their needs, monitoring the finances and bidding for work. Keeping their teams of engineers stretched, challenged and motivated is inevitably important for successful firms.

Many of the consultants, most of whom are members of the Association of Consulting Engineers, work internationally and have offices all over the world. Babtie Group, for example, employs 2,500 people in seven countries on projects such as the Singapore mass rapid transit (MRT) railway system, the Great Ankara Sewage system in Turkey and tunnelling for Railtrack in the UK. The work often involves quite a degree of travel. On larger projects a consultant will be on site much of the time but generally they will visit regularly to assess that progress is in line with the specification, within budget and on time.

The Association of Consulting Engineers is the trade organisation of these consultants and its Website contains details of its member firms.

CASE STUDY

Profile

Anthony Potter, civil engineer, Ove Arup Partnership, London

'After graduating in civil engineering I took a year out to go to a village in Ghana, West Africa, where I was involved in building a school, financed by the Japanese government. I improved the organisation of the works, ran the site and the materials stores and encouraged people to complete tasks in a sensible order. It was a great introduction to civil engineering.

'Back in the UK I joined Ove Arup in London where I work in their Infrastructure Division. My role is to work closely with architects to develop their plans. When the plans don't make good engineering sense in terms of design or cost it's my job to think of more appropriate concepts. I encourage environmentally friendly designs, particularly for the water, electricity and gas supplies, and I also ensure that developments won't flood. Some of my work is on the detailed design of water supplies. I calculate the quantities of water required and design systems to supply water throughout the site.

'I've been working on the Jubilee Plaza project at Canary Warf in London's docklands and Singapore Vista, a massive new science park in Singapore, both of which have provided some challenging and interesting problems.

'In the next year or two I am hoping to work on an overseas assignment for a disaster relief charity called RedR. Assignments last three months and involve setting up clean water supplies to refugee camps, preventing the outbreak of cholera through well-designed sanitation. It is an excellent use of practical skills that an engineering career can develop.'

Contractors

Civil engineering contractors turn drawings into reality. It is their job to take responsibility for a construction site, work out the levels, build the foundations and construct whatever new facility is required. Not only must they understand the civil engineering but also manage the contract. This involves scheduling the work of subcontractors who are involved with some part of the operation, ensuring that the right materials arrive on time and are used correctly.

Trainees are given a range of jobs including mapping the levels and setting out the site, supervising the pouring of concrete, working out how to overcome problems with water or drainage or difficulties with foundations. This may include taking soil samples for geo-technical analysis. Trainee engineers gradually gain responsibility for part of the site, for liaising with particular suppliers or pacifying local people who are inconvenienced.

Eventually, professionally qualified civil engineers are responsible for their own project and become site managers. At more senior levels their work includes bidding for contracts, recruiting and keeping teams of civil engineers together, liaising with clients and maintaining a watchful eye on personnel and costs. The largest contractors offer 'design and build' services to their clients and maintain their own design offices. Smaller ones rely on consultants or architects for this.

Civil engineers employed by contractors spend most of their time on site. They have to be mobile, since no one knows where the next job will be, though large contractors are organised into regions and small ones tend to only cover an area within 30 or 40 miles of their head office.

Contracting engineers have to be good managers of civil engineering situations, people and equipment, which differs from those employed in consulting. Engineering consultants have to be more creative, analytical thinkers with the ability to communicate their ideas to clients and contractors alike.

Major employers

Local authorities

Metropolitan authorities and county councils all have civil engineering departments. They are required to fulfil the council's responsibilities for roads, underpasses, drainage, street lighting, state schools and other civic buildings. Few councils now have contracting departments of their own, so much of the work involves deciding what needs to be done, appointing a contractor and supervising the construction. The work includes liaison with elected councillors, approving engineering drawings and plans and ensuring that the job is completed to an acceptable standard and within the budget. Most local authorities regularly recruit trainee engineers.

Other employers

Railways and water companies are important employers of civil engineers. As we have seen in the Hatfield rail disaster, maintenance of track to very high standards is essential. Bridges and tunnels, viaducts and cuttings also have to be maintained. Currently new lines are being laid from London to the Channel Tunnel and up the east of England from London to Edinburgh, providing safer and faster, more comfortable travel for the years ahead. Civil engineers are responsible for the design and implementation of all these initiatives. As Railtrack develops its stations to include commercial buildings, just like airports have done, their skills will also be applied in areas currently seen as unrelated to trains.

Water companies operate purification plant and have extensive distribution systems that get water to every home. During the 1990s they had serious problems with leakages, and civil engineers are responsible for the systems that distribute drinking water and collect and process sewage. It's a big job and many engineers employed in this area take additional postgraduate qualifications in public health engineering.

The Highways Agency is in charge of all the motorways and trunk roads throughout the country and employs civil engineers in the planning and construction of new roads as well as the supervision of maintenance and repair to existing ones. Some of this work is strategic, working out how to achieve set objectives, while time is also spent supervising contractors on site. On motorways the work often continues throughout the night and at times when traffic is slack. Systems have to be devised that will keep the traffic moving while operations continue.

Skills required

As we have seen, the job of a civil engineer varies in different contexts. Some engineers, especially those working for consultants, specialise in design and need to be creative and inventive. Others, employed by contractors, manage construction sites and have excellent organisational and management skills. All must be numerate, for engineering always involves numbers, but some need excellent mathematical skills to work out the stresses and strains that will arise within structures. Civil engineers employed by local authorities must be excellent communicators, ready to discuss projects with elected representatives, consultants, contractors and even the general public. Most need to be firm and confident when dealing with contractors or subcontractors to be sure that the construction progresses to the right specification, is high quality and within budget.

STRUCTURAL ENGINEERS

Structural engineers are the mathematicians of the construction business. (Quantity surveyors, who are not covered in this book, also use maths but to analyse costs.) It is their job to analyse the plans for new projects, consider how the weight will be distributed and where the stresses and strains will occur in the structure. This work involves taking each part of a structure in turn and analysing what loads will be placed on it and whether it is strong enough to sustain them.

A technique known as finite element analysis involves considering the stresses and strains that occur in each small part of the structure and building up a picture of what is happening overall. Having considered the loads of each part, eventually the entire structure is put together and analysed as a whole.

Studies

Many universities offer degree courses in civil and structural engineering that cover both subjects, and aspiring structural engineers work side by side with their civil engineering peers. A few universities, including Manchester, Sheffield, Edinburgh and Queen's Belfast offer courses in structural engineering and architecture. University College London and Heriot Watt University offer degrees in structural engineering. Students on these courses undertake the same studies as civil engineers for the first two years. These include structural mechanics, soil and geo-mechanics and the strength of materials. In the final year of these courses structural engineers take different studies in such areas as skeletal structures, continuum structures, finite element solutions and rock mechanics. They also complete a structural engineering project.

Major areas of work

Structural engineers working in consultancies often have to work on the designs of tall buildings and consider stresses that may be caused by wind. When oil installations are placed in the sea, problems that arise due to tides and waves also have to be taken into account. In some places buildings and roads have to be built to sustain the stresses that arise during earthquakes.

These engineers are not only concerned with the design of new buildings; they also have a role to play when existing buildings develop cracks. The leaning tower of Piza and the UK parliament's Big Ben tower are just two examples of buildings that have been saved from collapse by structural engineers. They examine the problem, develop models of the stresses that the building is subject to and work out ways of improving the situation.

Major employers

Structural engineers are not confined to working on buildings: their skills are also sought when aeroplanes, ships, turbines and other large structures are being designed. Firms such as Rolls-Royce that design turbines and aero engines are among the recruiters. The Ministry of Defence has large interests in ships and aircraft, rockets and satellites, all of which are subject to considerable stress in service. They and their contractors, such as Quinetiq, also offer jobs to these engineers.

Skills required

These engineers are office based. They are highly methodical with excellent attention to detail. Sound analytical skills and computer literacy are among their armoury of talents. They use mathematical models to investigate where stress occurs and work out how the structure can be altered to dissipate the problem.

Chapter 4
MECHANICAL ENGINEERING

Just about anything you touch that has been produced in a factory has had some attention from a mechanical engineer. Their contribution may be in the design, development and manufacture of the machinery on which it was made, or the design, production, testing and maintenance of the article itself. This gives mechanical engineers a very broad range of employment opportunities.

These engineers are responsible for fashioning materials into shapes and designing, developing and making equipment that will withstand vibration and thermal stress. Lubrication and wear are important concerns, especially when machinery rotates and vibrates as it does in an engine, pump or turbine. Tribology, the study of surfaces, is an area in which mechanical engineers are particularly interested, especially in what happens when two surfaces rub together.

Mechanical engineers are expert in forming and joining materials. They can use lathes and drills, welding, riveting and other techniques to get materials into the shapes they need, or make moulds into which fluids are injected that solidify to become shaped articles. Some engineers are taken on as apprentices at the craft level and may in time progress to professional engineers. Others join the industry after completing a degree course.

Studies

Engineering design is central to most studies of mechanical engineering because it illustrates and includes most facets of the subject. Students focus on topics such as the mechanics of solids and machines, materials, engineering drawing, thermodynamics and fluid mechanics. Subjects that tend to be included towards the latter half of mechanical engineering courses are power transmission, dynamics and vibration.

Some degree courses offer the opportunity to specialise, usually towards the end of these studies, though there are specialist degree courses in such subjects as naval architecture, manufacturing system engineering, process engineering and automotive engineering for those who already have a firm idea about their career direction. All degree courses present an initial broad foundation in the subject on which specialist expertise can be built.

Many mechanical items are often controlled electronically. This is especially true of robots and industrial machinery, weighing and measuring machines, valves in processing plant, lifts and escalators. Electronics is included as one of the subjects on a mechanical engineer's curriculum, not because these engineers need to be experts in electronics but so that they have a clear understanding of the electronic control systems that interface with mechanical devices.

Main areas of work

Consulting

A career in engineering consultancy attracts some engineers, as we can see later with Christopher Royle. In this capacity they may be designing the floodlight structures for a sports stadium or the mechanical handling equipment for a warehouse. Some specialise in building services such as heating, ventilation and refrigeration while others provide repair and maintenance services for complex machinery.

Contractors

Working as contractors, mechanical engineers are often commissioned to construct something such as a telecommunications mast or the pipework for a petrochemical plant. In that capacity they are responsible for the detailed design, the logistics of getting all the materials to the site and the construction and installation of whatever machinery or processing equipment is involved.

Industrial designers

So far we have not mentioned consumer goods such as kettles and irons, hair dryers, washing machines, garden equipment, bicycles, mobile phones, hi-fi equipment and computers. All the makers of these household goods and many more employ mechanical engineers.

You might think of a mobile phone and a hi-fi set as electronic but the casings and all the bits and pieces required to connect the cables are designed and made by mechanical engineers.

Industrial designers have the job of making things like irons, phones and television sets, prams and umbrellas so attractive that people will want to buy them. This type of design is a half-way house between art and engineering. For the creative, innovative engineer there are opportunities to be had there. Not only must the article be cleverly designed but careful thought has also to be given to the machinery on which it will be made.

Medical applications of mechanical engineering include metal hip joints and other body implants plus body scanning machinery and pace-makers. Smith and Nephew are among those who make metal structures to help support people who experience problems with their spine. Engineers engaged in this area often choose to be members of the Institution of Physics and Engineering in Medicine rather than the Institution of Mechanical Engineers, which attracts the majority.

As an engineering technician you could be employed in a workshop to fabricate a part of some larger structure. You might supervise a production machine that is programmed to drill holes or fashion metal into a certain shape. When such processes are automatically controlled they are called computer-assisted engineering (CAE). Your job might include the maintenance of a piece of equipment – we all know our friendly car technician whose detailed training is often arranged by the car manufacturer.

As an incorporated engineer you could be designing a product or supervising a workshop, responsible for maintenance and repair. You could be working in many of the jobs listed in Chapter 1 applying well-tried technology to solve problems.

Chartered mechanical engineers are more often responsible for novel projects. They might be pushing the frontiers of technology, working in research and development or acting as a consultant engineer.

Major employers

Employers of mechanical engineers cover a very broad range and you can join them at many different levels, from technician to chartered engineer, depending on your qualifications.

In the aerospace industry these engineers are working on the assembly of planes and helicopters, rockets and satellites. Aeroplane parts are often manufactured in many different places and transported to a site where the final product is assembled. Airbus, Rolls-Royce, BAe and GKN are among the employers. Hydraulic systems for lifting undercarriages and moving flaps on the wings of aircraft are important aspects in which mechanical engineers are involved.

The automobile industry relies heavily on these engineers, sometimes called automotive engineers to indicate their specialist focus. This vast international industry produces a host of different products, from taxis and buses to tractors and tanks in addition to the millions of cars of every shape, size and colour that come off its production lines. Mechanical engineers work in all departments, from designing the engines and power transmission to the wheels, known as the 'powertrain', to the manufacture and maintenance of production equipment. They manage the production lines and thoroughly test the final products to ensure that quality and safety are maintained.

CASE STUDY

Profile

Christopher Royle, mechanical engineer, Ove Arup and Partners, Nottingham

'I'm working for Ove Arup in automotive consultancy on projects for clients who are car manufacturers. It involves looking at what happens to vehicles when they are involved in a crash and devising systems that will optimise the protection of passengers.

'I analyse the energy involved in the impact and the way that it is absorbed by the structure in a number of different front, side and rear incidents, against legislative and consumer-driven requirements. The car occupants are better protected if the vehicle crumples smoothly over as long a period of time as possible. Some of my work also aims to improve the in-car protection systems such as seat belts and air bags and how they operate during a crash.

'I graduated in mechanical engineering at Nottingham University and spent some time in industry researching hot, high pressure piping systems in chemical plant. I redesigned them to reduce the stress they work under in order to reduce the risk of leakage and critical failure.

'Then I returned to university to research for a PhD developing a low-cost miniature sensor for vehicles that measured the rate of turn (angular velocity) utilising the principles of gyroscopic motion. The system I designed has many potential applications including active suspension for improved handling and in satellite navigation systems.

'Now a consultant at Arups I'm training to become a chartered engineer with the Institution of Mechanical Engineers.'

Before leaving transport we should mention that these engineers have a role in the production of ships and submarines, train engines and carriages, though work on ships in the UK is now almost entirely confined to refitting and upgrading. The railways offer numerous challenges including the installation, maintenance and testing of track, points, the structures that carry signals and the gantries for overhead power lines.

CASE STUDY

Profile

Belinda Okpere, trainee signalling engineer, Railtrack

'I graduated from Brunel University with a BEng in mechanical engineering and now I'm training to become a signalling engineer with Railtrack. I've had several placements. Currently I'm working on designing and developing circuits for signals at particular railway locations. Previously I spent some time in maintenance and installation, checking equipment on site. This included the vital question of signal siting – are signals in the right place and can they be seen by drivers? I went out onto the railway, investigated the signal sites and made recommendations.

'Railtrack is sponsoring me on a MSc in Engineering Asset Management at Robert Gordon University. I spend one week a month at the university attending lectures and completing practical work. It's an intensive 40-hour week and I will be completing a project that relates strongly to a Railtrack problem.

'When I'm a fully trained signal maintenance engineer I will be developing strategies for maintenance and making sure that we get the best out of our systems. It will involve asking questions such as why do we do it this way and can we do it better? What shall we change and when is the best time to do it? I will be supervising contractors to make sure that they are completing their work in line with the contract and to the required standard. It's a challenge and very rewarding to be continually improving our rail transport system.'

Continuous industrial production nearly always depends on transporting items around a factory, lifting, turning, mixing (liquids) and using conveyor belts. Whether the equipment is making food or paint, cars or television sets, these operations are the province of mechanical engineers. Wrapping and packaging processes are also their domain.

The oil and gas industry is another important employer. Most of the non-electrical equipment used in the exploration, production, transport and processing of oil and its products is provided and used by mechanical engineers. They can be working as drilling engineers, pipeline engineers, designing, installing or maintaining pumping stations, or working on equipment for tankers to use in the docks. Technicians working with drilling engineers are known as 'roughnecks'. Mechanical Engineers also build large storage tanks and refrigeration systems that liquefy natural gas under pressure.

Skills required

Mechanical engineers are highly practical people with a good eye for dimension. They can visualise how an item will be, how it will fit in with other parts of an assembly. Engineering drawings are used extensively. Because mechanical things are used in virtually all other aspects of engineering they have to liaise with and work in project teams with engineers of most other disciplines, so they must have a general understanding of those and an ability to see where other technologies interface with theirs.

Chapter 5
ELECTRONIC AND ELECTRICAL ENGINEERING

Modern living is utterly dependent on electricity and electronics. Mobile phones, computers, cash machines, security systems, DVDs, hairdryers and much more are all based on this technology. It's an area that moves on every year with innovations and fast-changing technologies, so naturally it provides employment for large numbers of engineers.

ELECTRONIC ENGINEERS

The world of electronics is moving from analogue to digital systems and from situations in which machines are connected by wires to those where they contact each other via radio links. Optical laser driven systems that can take much more information down cable of the same diameter are rapidly replacing other methods of electrical communication. At the centre of it all is the integrated circuit, a vast array of electronic circuitry that can be accurately produced on a small piece of silicon. No wonder that a career in electronics is such an attractive proposition.

Studies

Electronic engineers study such subjects as circuit theory, fields and waves and digital electronics. They learn about microprocessors and electronic devices and all the components that make up electronic systems. They become conversant with how electronic signals are transmitted, received and processed. Different courses cover different areas of the subject. These might include telecommunications, microwave engineering, electronic networks and opto-electronics. Design projects and laboratory work are often used to provide the practical experience that applies what they have learnt and enhances their understanding.

Main areas of work

Electronic engineers have many choices in the career path they can take. Some concentrate on integrated circuits, semi-conductors and electronic devices. Others specialise in developing control systems for machines, or navigation systems for aircraft and ships.

Telecommunications

In recent years telecommunications has been an area of strong growth where engineers have been developing mobile phone systems and installing transmission sites just about everywhere. Getting electronic circuits into phones of ever decreasing size without causing electrical interference between the different components has been a major challenge. Radio frequency and microwave engineering is still an area where there is a shortage of people who have the requisite expertise.

The range of career opportunities available for electronic engineers is extensive. These engineers often work on the design and development of new electronic systems or the modification and updating of existing ones. A few work in research investigating novel ways of doing things, new methods of circuit design and ways of distributing more information down the same wires. Discovering different approaches to the problem of an ever-increasing need for the processing and distribution of information is often part of the task.

There has been a great deal of research recently on fourth-generation mobile phones and methods of passing information from one piece of equipment to another without wires.

Employers include equipment providers such as Siemens, Nortel Networks, Nokia and telecommunications companies like BT and Vodaphone.

CASE STUDY

Profile

Richard Day, electronic engineer, Siemens

'I graduated from Nottingham Trent with a degree (BEng) in electrical and electronic engineering and followed that with a MSc in electrical engineering at Nottingham University.

'Now I'm working for Siemens at their Roke Manor Research Centre. Engineers begin here as graduate engineers and get promoted through the ranks to engineer, senior engineer, group leader and consultant.

'I'm a team leader with a group of five engineers working on systems for testing node controllers for the third-generation base stations required for telecommunication systems. We are designing an in-house terabit router for the Internet protocol and wrote the specification in a high-level language.

'My previous job was the design and testing of an integrated circuit on a silicon chip for a specific application. It was a £50 million product, so a lot of responsibility. Integrated circuits are extremely complex and working out systems to test them is quite demanding. We use C, C++ and VX works programming languages to complete the software we need to get the data from the chip.'

Work in research and development for a large organisation in the industry and you could be in a team of 20 engineers tackling a major project. Decisions will be taken at the outset on which functions will be achieved electronically and which using software. Electronic engineers work closely with software engineers in these circumstances to integrate each other's design into the final product. In smaller organisations project teams may consist of just two engineers, and electronic engineers who also possess strong software skills are particularly attractive.

Defence

Defence engineering is a key area of employment where secure communications, command and control systems, navigation, guidance systems for weapons and other state-of-the-art security systems are continually being developed, produced, tested and maintained. The theatre of war can be in hot sandy countries or cold icy ones: whatever the conditions the equipment must work effectively first time. Leading employers include Qinetiq, BAe Systems and Marconi.

Control

Electronic control systems are used to operate industrial plant and processing equipment. Oil wells way out at sea, distribution systems for electricity, water and gas supplies are all electronically controlled. The Dockland Light Railway in London operates without train drivers and serves as a typical example of what control systems can achieve. Sometimes an element of artificial intelligence is included that allows the system to make decisions using software that accesses information in a series of databases. ABB is the leading supplier of control equipment but all companies in the manufacturing and processing industry recruit control engineers.

Oil exploration

The oil industry is an important employer because it uses electronics not only to control its industrial processes but also in the exploration for oil. Seismic and magnetic surveys all use sophisticated electronic equipment and, once an oil well has been drilled, electronic tools are lowered down the hole to investigate the structure of the rock and detect the presence of oil. It's an international business attractive to those who are keen to travel to far-away places that are often inhospitable and cut off from society. Recruiters include BP, Shell, ExxonMobil, TotalfinaElf and ENI the Italian oil company. Service providers such as Schlumberer are also important employers in this sector.

Robotics

We have already mentioned industrial control systems. Not only are manufacturing lines automated but they also use robots. A robot is a multi-disciplinary engineering achievement drawing on the design skills of electronic, mechanical and software engineers. It can go into hazardous environments such as nuclear reactors, complete repetitive tasks like welding and spraying, or move equipment around to order. Employers include the users of robots, especially the car and electronics companies such as Ford, Nissan and Marconi.

Skills required

Electronic engineers need a good understanding of physics and maths and the ability to apply it to their work. As designers they have to be innovative and creative. On major projects they usually work in teams and need excellent team skills, though occasionally they will work alone and have to be self-reliant. The ability to communicate ideas verbally and through drawings and written reports is another essential skill.

Often they work with software engineers on designs of systems that include both software and electronics. They also get involved in mechatronic systems that are at the interface between electronics and mechanical engineering. In these circumstances it is essential to have some understanding of the relevant technologies and to be able to work closely with engineers from other disciplines.

ELECTRICAL ENGINEERS

Electrical engineering is a much older subject that electronics, involving heavy current and high voltages. Electricity is essential for heating, lighting and providing power to production machinery, lifts, cranes and trains.

Studies

Most of the HND and degree courses available in the UK include some studies in electrical engineering. Students study subjects that include energy distribution and utilisation, electrical power engineering, waves and fields, electrical machines and drives. In some courses they also investigate safety, reliability and engineering management.

Major employers

Electrical engineers work in a range of areas. In power generation the electricity companies employ them in the generation and distribution of electrical power. Many factories have their own power supply and need engineers to install and maintain equipment. Hospitals all have to maintain independent power supplies just in case they are interrupted. These engineers are also employed by the National Grid on the distribution of electrical power around the UK.

Another side of this business is the firms which make turbines for the generation of electrical power; they also need to employ electrical engineers. Rolls-Royce, for example, produces turbines for large and small power stations world-wide. Engineers are not only concerned with their design, development and manufacture but also provide an after-sales maintenance and repair service.

All electric railways, including the London Underground and Railtrack need electrical engineers to keep their services going smoothly. New systems, such as the railway between terminals at Gatwick Airport rely on magnetic levitation, a system devised by electrical engineers that reduces friction and wear on the track and train.

Manufacturing companies need to supply power to their machinery, especially when there is heating involved or a need for heavy machinery, belts or lifts to move large items. Electrical engineers are employed for this purpose.

Engineering consultants working in the area of building services also employ electrical engineers to develop lighting systems, heating and ventilation and sometimes refrigeration. The technology may be old and well developed but the opportunities to make a career in this field are still strong.

Professional bodies that look after the interests of these engineers include the Institution of Electrical Engineers, the Institute of Measurement and Control and the Institution of Incorporated Engineers.

Skills required

When dealing with high currents and voltages, safety is of the utmost priority, otherwise people could be exposed to electrical shock or fires started through sparks from electrical short circuits. These engineers often use colours to identify particular leads and the resistance of a resistor, so colour blindness is not generally acceptable. A high degree of care and attention to detail is required.

Some of these engineers work outdoors and may need to travel to installations by car, so a driving licence can be important. The ability to interpret drawings and discuss them with technologists from other disciplines is also a significant advantage.

Chapter 6
MATERIALS ENGINEERING

Choosing the right materials for each particular use is an essential part of any engineering project. Materials engineers, sometimes called materials scientists, have an intimate knowledge of the properties of a vast range of materials and how to treat them to change those properties for any particular application. They study metals, ceramics, plastics and rubber, glass, paper and textiles. In some cases their education starts with a general course covering the broad range. In others they immediately specialise in one area such as metallurgy, electronic materials or textiles.

Diffusing one material into another, producing alloys or composite materials are all a part of their armoury. Coating materials to make them harder or tougher and aligning the grains within the material to introduce alternative properties in different directions are also among the methods they can use. Surfaces can be treated to improve their performance in certain situations and reduce their susceptibility to corrosion.

Integrated circuits used in sophisticated electronic equipment such as mobile phones include crystals of silicon or germanium on which are deposited a series of layers in different configurations, often less than a micron thick.

Studies

When studying, these engineers are focused on subjects that include the production, structure and properties of materials. Most cover the entire range – metals and alloys, ceramics, glass, polymers and bio-materials. A few specialise in areas such as aerospace materials, electrical materials and bio-materials, including tissue engineering and bio-compatibility.

Studies of production include how materials are processed and formed, including molton metal flowing into moulds, being extruded into wire or rolled to make sheet. Plastics are often injection moulded or pulled into thread.

At the outset students study the theoretical principles of materials science and engineering. This includes investigating thermodynamics and mechanics, deformation and fracture, the physical and chemical structure of materials and manufacturing processes. Students learn the broad range of techniques for examining and testing different materials.

These courses include a lot of laboratory work, discovering in practice what has been discussed in lectures and tutorials. A design project is usually completed in the final year that usually involves making a component for an engineered product.

In our environmentally conscious society many materials are recycled. Metals, plastics and paper already used and collected are often introduced into the production process. On some courses students learn how recycled materials can be blended with new ones to manufacture products.

Main areas of work

In addition to organisations that produce materials, or use materials to make other products, defence is a key sector. Materials used in defence equipment from rockets and torpedoes through to ammunition, communications systems and clothing all have to withstand extreme conditions and be part of machinery that always works first time. In rocket motors, for example, power comes from an explosive charge placed within a metal cavity. The metal and the rest of the structure must be designed to withstand the extreme stresses that occur at lift off.

The energy sector is another important employer. Oil exploration places big demands on the materials involved. Installations placed in the sea must withstand salt water, waves and wind, while the tips of drills used to drill wells must grind their way through all manner of geological structures and work well under high temperatures and pressures. Pipelines have to resist corrosion, and petrochemical processing plant must work at high temperatures and pressures in a corrosive environment.

When we look at renewable energy such as wind and solar power, again the materials employed have to be well designed. Solar panels must let in the energy but not release it, so the optical and thermal properties of the materials used are of the utmost importance.

We live in an age where packaging is given a high value. Food containers must be tamperproof, drinks packages must not corrode or leak, and produce has to be maintained in prime condition until we are ready to use it. All these needs present materials problems. Some foods such as chips and crisps are manufactured into shapes, and then the potato, meat or other food is extruded or processed in some other way just like a metal or plastic.

The most sophisticated uses of materials are in sensors. These devices detect parameters such as stress and stain, temperature changes, humidity and other physical dimensions. To make a sensor, different materials must be put together in such a way that they react sensitively to a change in the parameter that is being monitored. It's often a challenging task but one taken on by materials engineers.

Materials engineers have a variety of roles. Some focus on design and the optimum choice and use of materials within any given product. Others work in production, either manufacturing the materials or products made with them. Many research and develop new materials, or set standards for safety or labelling of materials.

When materials fail there is usually an analysis of what happened and why. This is another key area of employment in which these engineers use a range of tests to diagnose what went wrong and so avoid a repetition in the future. They include optical and electron microscopy and tests for toughness, hardness, creep and other properties.

A small area in which these engineers work is forensic science. They may have to analyse materials that were left at the scene of a crime and establish where they came from. Alternatively they are often asked to decide why something happened. Why did the rail fail causing the crash at Hatfield? What is the likelihood that it will happen again? When a car crashes, a ship goes down or a chemical plant explodes, is it the result of a failure in the materials involved and if so who was responsible? This is a question insurance companies often need to answer. Materials engineers investigate.

Major employers

Employers of materials engineers fall into two major categories. First are the producers of materials, including Corus and Pilkington, the makers of steel and glass, which manufacture their products in a range of forms and advise customers on the best ways in which to use them. Second are those companies that take the materials and use them to make other products. This group includes car manufacturers, producers of turbines, aero engines and pumps where vibration and heat distribution are key factors in the performance of the materials involved. The electrical devices sector, including makers of optical fibres, wires, integrated circuits and semi-conductors, also employ these engineers.

In the aerospace industry having materials you can rely on is of paramount importance. Thirty years ago Rolls-Royce went bankrupt because it relied on a new composite material for its aero-engines that failed in service. This is probably the largest industrial problem ever caused by using materials that were unsuitable for their purpose.

CASE STUDY

Profile

Mustapha Rampuri

'I graduated in materials design and processing from Nottingham University and took a job as a graduate engineer with BAe Systems. My interest in chemistry had always been strong but I wanted to make new things. The idea of being stuck in a lab mixing chemicals all day didn't appeal and that's why I decided to go into engineering.

'During my degree course I worked for three months at Powergen researching the submerged arc welding of turbine rotor discs. They wear out with use and ways have to be found to fix them rather than replacing them too often with new ones. It was a great experience because it gave me some commercial awareness that I had lacked before.

'At BAe Systems I'm not assigned to any particular area, so I can carve out my own career path. For the first five months I was working on the materials in rocket motors. It was a £3 million project that involved the design and assembly of a manufacturing facility for rocket motors. Eventually I became the project manager with responsibility for its day-to-day running. It was daunting at first. I have to be on the ball because I am managing staff, some of whom have been here 25 years, but it's developed my management skills and in engineering there is always someone around who will help.'

Most industries have research and technology organisations attached to them that carry out contract research. EA Technology provides such services for the electricity industry and RAPRA does the same for the rubber and plastics industry. There are around 40 such organisations, all members of the Association of Industrial Research and Technology Organisations (AIRTO) and most of them are significant employers of materials engineers.

Choose materials engineering and you have a diverse range of career opportunities.

Skills required

Materials engineers are meticulous in their approach to analysing problems. They need excellent laboratory skills in performing investigations. Usually they are performing a service for other engineers, who are using the materials for a range of applications, so they need to be good communicators and team players.

Chapter 7
CHEMICAL ENGINEERING

Chemical engineers have a key role in a range of industries in which chemicals are processed. They include the oil and plastics, agro-chemicals, pharmaceutical and biotechnology industries. The processing of food and drink, soap and hygiene products, perfume and explosives is an area in which chemical engineers have an important role to play.

Biochemical engineering is the application of chemical engineering principles and techniques to the processing of living organisms. These engineers specialise in the growing technology of making products using genetic material. It's a technology that in some senses is thousands of years old – the fermentation of wine and the use of yeast in bread making being two early examples. Given our dramatically increasing knowledge of DNA, genetics, proteins and enzymes, it is now moving forward in leaps and bounds. Most of our medicines and diagnostic kits are likely to be produced biochemically in the future.

Studies

During their studies chemical engineers develop a basic understanding of chemistry, microbiology and mathematics, the key subjects that underpin their own discipline. Students of the subject develop an understanding of the kinetics of chemical reactions, the dynamics of fluids and how heat is generated and transferred around a chemical plant. They study the molecular basis of product and processing engineering from which everything else ensues, including the principles of processing plant and the different types of unit that are included.

What goes on inside a chemical plant includes the use of membranes and phenomena related to how the surfaces of liquids and powders behave. Chemical adsorption and ion exchange are among the subjects that these engineers have clearly to understand.

In biotechnology students learn about the development of microbial strains and the structure of macromolecules. Reactors in which the biological reactions take place have to be biologically clean and their design requirements differ from those of normal chemical plant.

There is a considerable quantity of laboratory work that includes the use of pilot plant facilities to carry out extended experiments. All modern chemical processing plant includes sophisticated control systems that require software and electronics. Chemical engineers have to make themselves conversant with these. They also use modelling and simulation techniques to help them to predict what will happen in different circumstances, temperatures and pressures when chemical reactions are taking place.

Main areas of work

The chemical industry is a business and engineers are intimately involved with such issues as the economics of processing. Are there more cost-effective ways to reach the same solution? Can a product be made using cheaper ingredients? What are the safety issues surrounding specific designs and how can these be addressed?

Environment

Chemical engineers are very concerned about what happens to waste products. They prefer to be seen as professionals who preserve and protect the environment rather than as engineers who cause environmental problems. Trainees learn to decide which effluent chemicals must be treated to transform them into acceptable waste products that can be discharged into the environment and how this can be achieved. Effluent is continually sampled, monitored and measured to ensure compliance with set standards and regulations.

The oil industry

The oil industry is a key employer of chemical engineers. At the oil well chemical engineers are concerned with the flow of oil and its separation from the gas and water that emerges. Petrochemical plant is managed, maintained and upgraded by chemical engineers. It turns the crude oil into a range of products, from lubricating oil and tar for the roads to plastics and chemicals to be used as the starting point for other processes.

CASE STUDY

Profile

Suzanne MacDonald, chemical engineer, Fife ethylne plant, ExxonMobil

'I joined ExxonMobil straight after my degree in chemical and process engineering at Strathclyde University and became a process engineer. As a graduate trainee I was given experience in a range of different functions. It included designing new equipment, test work on the plant, computer modelling and energy conservation projects.

'I designed a "blow-down drum" which cools waste water and steam before it goes into the effluent stream. The one that had been used previously was corroded so I used a new material.

'On another project I tested filters that remove oil from the water and set up a pilot test rig running a slip-stream from the plant. I changed the configuration and size of the filters to increase their effectiveness.

'Now I'm a business analyst and my job is to help write the corporate plan. It involves setting production targets for next year, examining costs and setting budgets. I look at how costs might be reduced and performance targets met. Every month I have to present a report to the European management, so it's quite a challenge.'

Gases

Chemical engineers working in this environment are developing new plant to meet customers' ongoing needs, having a sensitivity to the costs involved. They can be discussing with clients precisely what their requirements are and installing and maintaining gas production equipment on site.

Food and hygiene products

As a chemical engineer in this environment you could be running a detergent factory, developing novel processes or visiting food manufacturing plant to assess the effluent. Alternatively you might be responsible for optimising the productivity and reducing the quantity of waste products emerging from the plant.

Petrochemical contractors

Having agreed with their client the specification, the quantities of which chemicals the plant will produce and the levels of purity required, petrochemical contractors organise construction on site of the processing plant, the installation of all equipment and the commissioning to full operation.

Having started their career in the design department and progressed to managing contractors on site, engineers working in this area can also be involved in the planning, logistics and scheduling of the work and ensuring that all the materials arrive in the right place on time. At a senior level they are bidding for new contracts, preparing and costing proposals for future work and developing and maintaining their teams of engineers.

Water

The water we drink is one of the purest products that are supplied to the public. Chemical engineers have a key role to play in the production and distribution of drinking water. They also work on the systems used in the processing of wastewater and sewage where biochemical degradation techniques are in use.

Major employers

BP says that its chemical engineers may be involved in 'process design and the optimisation of plant performance, technical development or specialist technical roles. You could be working on projects as diverse as manufacturing and production support, the development of innovative processes or technical support for the application of new and existing products.'

Some of its engineers also have jobs working in the area of health, safety and the environment. They visit plants and take samples of air and effluent. Analysing the risks and protecting both employees and local residents from harmful pollution is central to their work.

Air Products and BOC are firms that provide gases, mainly oxygen and nitrogen for customers who need them for their own production plant. They also work on cryogenic systems to liquefy natural gases.

Firms such as Unilever and Procter and Gamble, Boots and Rhodia Consumer Specialties Ltd manufacture foods, soaps, perfumes and related products.

Petrochemical contractors such as Fluor Daniel and Bechtel employ engineers to work on the detailed design of processing plant that may be required in any part of the world.

WRc (the water research centre) is the leading research and technology organisation in water supply, employing many chemical engineers, but the water companies also seek their services.

Skills required

Employers of chemical engineers don't want to recruit people who just have a good understanding of engineering. They're also looking for engineers who can continually accept new challenges and focus on solving problems and getting results. Self-confident people, who work well in teams, can see things from other people's perspectives and have the ability to convince others about the best course of action, are particularly sought by industrial firms.

Many of the degree courses for chemical engineers emphasise the personal development that is necessary if you want to become a professional engineer and provide an environment in which these qualities can be enhanced, in addition to the learning of technical skills.

Chapter 8
AERONAUTICAL ENGINEERING

Whenever aircraft are involved aeronautical engineers are never far away. They might be working on novel designs, simulating a range of situations in which an aircraft can find itself, or using a wind tunnel to discover how particular shapes of aircraft respond in certain situations. But as we shall discover later, these engineers often work in multi-disciplinary teams with other engineers and are not confined to the aerospace industry.

The key to success in this highly competitive industry is meeting customer needs. Some jet fighters travel at twice the speed of sound (Mach 2), and have wings that swing in and out. Others are designed for vertical take off and can operate easily from a ship or other conditions where runway facilities are simply not available.

In the civil aviation industry some planes are designed for short-haul domestic flights and others for intercontinental travel. Their designs differ because the payload (weight), size and fuel consumption vary to meet different operational requirements. It's a very diverse business where engineers work in teams to design and manufacture aircraft, or parts of aeroplanes, and also to maintain them cost-effectively.

Studies

Students of aeronautics take subjects that include aerodynamics, aircraft structures, aircraft operations and flight mechanics. All aircraft include hydraulic, electronic and navigational systems in addition to their engines and means of propulsion. Studies of these systems are an essential part of all aeronautical engineering courses. Aspiring engineers start from basics by considering the materials that are used in aerospace systems and the stress, heat and pressures that the aeroplane structure will have to withstand during a flight.

Discovering solutions to the problems that arise requires a strong knowledge of mathematics and an ability to apply it in complex situations in order to reach workable solutions. Safety is of paramount importance so attention to detail is an essential quality.

Main areas of work

Systems engineers

Systems engineers analyse customers' needs and translate those into products. This involves specifying technical requirements and developing designs using computer-based modelling. They plan and conduct trials to test their designs. Those working in research and development roles are continually seeking to create advantageous technological advances to stay at the leading edge of ever-advancing technology.

Aircraft maintenance

Commercial airlines and the Armed Forces all have to maintain their aircraft to high standards that ensure safety for their crew and passengers. This maintenance is a regulated activity and lengthy training is required of those who wish to become an aircraft maintenance engineer. Some work for airlines to inspect aircraft between flights, an activity known as 'line maintenance'. Airlines often outsource the maintenance function to specialist companies that provide this service and are also important employers. It involves a specified set of pre-flight checks, diagnosing faults and repairing the aircraft's avionics (eg navigation) or mechanical systems. Engineers usually specialise in one of these two.

When an aeroplane cannot be repaired on the tarmac, or it reaches the time for a regular overhaul, it has to be taken in for 'base maintenance'.

Other areas

Outside the aircraft industry, engineers are often involved where designs include large structures, such as the hulls of ships, and the optimum shapes for torpedoes, rockets and masts.

Major employers

The industry employs around 150,000 people in the UK and enjoys an annual turnover exceeding £18.5 billion. The Society of British

Aerospace Companies includes more than 170 firms each with a part to play, and the Royal Aeronautical Society lists more than 100 companies in its 'Associated Companies Scheme'.

At BAe Systems one of its activities is building wings for Airbus Industries, the trans-European civil aviation manufacturer based in Toulouse and Bristol. They must be 'aero-elastic' in their response to gusts of air and vertical loading. It's just one of the projects on which aeronautical engineers are engaged.

Aeronautical engineers work in many of its departments, especially engineering, systems design, software, customer support, procurement and marketing. In engineering they are concerned with the air-worthiness of planes, aerodynamics and the testing of equipment and structures, including flight tests.

CASE STUDY

Profile

Ewan Barr, aeronautical engineer

'I graduated in aeronautical engineering from Imperial College three years ago and now I'm working at Airbus in Bristol. My job title is Capability Development Engineer and what I do is concerned with everything that happens when an Airbus aircraft needs a repair to its wings. My job is to develop the tools and processes that are required by the aircraft repair team in Toulouse. They may need the services of our manufacturing department or our process development team. Specialist information may be required, so I organise that they can gain speedy access to data that meet their needs. They may need additional IT systems or more staff, so I liaise on their behalf with our IT and human resources departments.

'The process leads to a newly designed repair that it safe to go on the aircraft and return it to an airworthy condition.'

All companies recruit at several different levels. Westland, the helicopter maker, for example, offers engineering apprenticeships, technician apprenticeships, undergraduate placements and jobs for graduate engineers. Engineering apprentices must have five GCSEs, or their equivalent, at grades C or above, including maths and English. They train to become aircraft fitters, mechanical fitters or composite/mechanical engineers. This involves studying the NVQ level 2 in engineering manufacture, which covers electrical and electronic fitting, sheet steel manipulation, joining and welding. After successfully competing this course they can move up to NVQ level 3 and following this spend a year in one of Westland's specialist areas.

Technician apprentices must have higher qualification – five GCSEs at grade B and three at grade C. They also complete NVQ levels 1 and 2 but then go on to the Higher National Certificate in Engineering (part-time study) and spend a year in a department such as design or avionics.

Skills required

Aircraft maintenance engineers train to obtain a licence, awarded by the Civil Aviation Authority to maintain airframes, avionics, radio and engines. The academic requirement is usually four or five GCSEs at grade C or higher, including maths, science and English. Some employers seek higher qualifications including one or even two A-levels in potential recruits, who often join as apprentices. They attend courses that lead to qualification as a licensed aircraft maintenance engineer. These engineers are often members of the Association of Licensed Aircraft Engineers.

Currently there is a dire shortage of licensed aircraft engineers and new foundation degree courses in the subject at Kingston University and City of Bristol College are addressing this issue by providing all the relevant training. They are accredited by the Royal Aeronautical Society.

All the armed forces operate aircraft, not just the RAF, and all train their staff to manage and achieve a high state of readiness. The Ministry of Defence and its suppliers, including Qinetiq, previously the Defence Research and Evaluation Agency, are key employers.

Safety is of paramount importance whenever anyone is flying. Key qualities of those involved with aircraft are a thorough understanding of the technologies involved in aeronautical engineering, excellent mathematical abilities, attention to detail, perseverance and good problem solving skills.

Chapter 9
BECOMING A PROFESSIONAL ENGINEER

Let's come clean right at the start. While most engineers have academic qualifications, by no means all engineers have professional qualifications. Some employers don't encourage their engineers to take the trouble to get the right experience and training to become a fully qualified professional.

Many engineers do not bother to join an engineering institution but for those that do there are considerable advantages, especially if their work leads them into international projects. Gaining professional qualifications can definitely help your career. Some companies, especially global organisations such as engineering consultancies and oil companies, often insist that the engineers working on their projects are professionally qualified. Certainly most consultant engineers need to gain the highest possible professional qualifications because their CVs are closely assessed before each client engages them on a new project.

In the UK the Engineering Council (UK) sets the standards of professional engineering qualifications. They are all detailed in its policy document 'Standards and Routes to Registration (SARTOR)'. Engineers qualify by joining an engineering institution, often at first as a student member and going on to obtain the education, training and experience the institution requires before it will put them forward to register.

If you are studying engineering we strongly recommend becoming a student member of an engineering institution that is relevant to your studies. Membership is often free or heavily discounted. It gives you access to a whole range of resources that you would not otherwise be able to benefit from. These include use of the members' part of the institution's Website, which often carries job opportunities. But more than that you will receive regular publications giving you details of what is going on in your eventual profession. Some institutions organise events specifically for young members and they provide the chance to meet like-minded individuals who are engaged in similar situations. Building up a network of contacts and people who can help you discover opportunities as well as those with whom you might exchange information, is often a key to career success.

Before they can register for one of the professional qualifications engineers must have a relevant education followed by professional training and development that prepares them for a particular role in their chosen area of engineering.

We mentioned in the Introduction that there are several levels of seniority in engineering careers. First are the operatives who are responsible for a piece of equipment, then there are the craftspeople who can make an engineered item to a given specification and drawing. Their role includes fitting parts together or installing electrical wiring.

The three professional grades of the profession, however, are engineering technician, incorporated engineer and chartered engineer.

Whatever your goal in terms of professional qualifications you are expected to keep a detailed log of your training and 'on the job' experiences during your initial training. It includes a period of induction into your employer's organisation, and an introduction to the broad range of techniques that are used in that particular environment. Many of the engineering institutions accredit the training of firms that recruit engineers in their discipline. These are often listed on their websites. Leading organisations in this field offer a structured training scheme in which each trainee has a mentor whose responsibility is to offer help, encouragement and advice that will lead to successful career development.

INCORPORATED ENGINEERS

Incorporated engineers form the mainstream of professional engineering practitioners. They are highly qualified engineers who have the ability to apply and manage existing and developing technology. They have a detailed understanding of their area of engineering and may well be managers of facilities or key operational functions. Large pieces of complex equipment could be under their control. They may supervise inspection and testing procedures or develop diagnostic systems, but they are not expected to take quite as much responsibility or be as creative and inventive as chartered engineers.

These engineers have to experience a period of initial professional development to build their competencies through training and work experience in the area of engineering they are working in. The minimum academic qualification for incorporated engineers is a bachelor of engineering degree (BEng) or a Higher National Diploma plus additional studies designed to reach the BEng level.

The Institution of Incorporated Engineers is the leading institution for incorporated engineers.

ENGINEERING TECHNICIANS

The role of engineering technicians is to apply proven techniques and procedures to solve practical engineering problems. Under the guidance of incorporated or chartered engineers they make valuable contributions to a design, a manufacturing system, and the operation, maintenance and testing of products.

In common with the other grades they are expected to undergo a period of induction and professional development during which they familiarise themselves with their role and the business environment. The minimum qualifications required for engineering technician registration is the Vocational Certificate of Education (previously the advanced GNVQ) or its equivalent, such as two A-levels. Engineering technicians usually gain their experience through modern apprenticeship schemes (see example on pages 65).

CHARTERED ENGINEERS

Chartered engineers are most often the ones in charge of an engineering project. They are expected to work creatively and innovatively, developing and applying new technologies whenever appropriate. They are expected to have the ability to pioneer new engineering services, develop more efficient production techniques or improved designs.

When judgements are required about the level of risk involved in deciding between a range of options, chartered engineers are expected to be able to take the initiative. Some of the engineering institutions, including the Institution of Electrical Engineers, only cater for chartered engineers (see the list of institutions in the Useful Addresses section) while others have a broader remit.

The ideal educational route to chartered engineer status is to take a four-year Master's degree (longer if a sandwich or part-time course) that is accredited for that purpose by one of the engineering institutions. These degrees not only provide a broad basic education in engineering and specialist learning in a chosen engineering discipline; they also place engineering in its international context and give students an understanding of the wider business implications of their work. Design studies are always central to such courses and provide an 'integrating theme' through which many other aspects of the subject, especially the choice of materials and methods of production, safety, costing and marketing can be addressed.

In many universities students with good A-level scores or their equivalent, register for the MEng and take a decision after they have completed their first two years of study whether they will proceed to a Bachelor or Master's degree. In general only those in the top half of each year group – those considered able to gain an upper second class honours degree – are offered the opportunity to proceed to the MEng.

Fortunately there are other routes which aspiring chartered engineers can take in order to meet the educational requirements. One is to complete a BEng honours degree and then take further studies, which the Engineering Council calls a 'matching section' to bring their academic expertise up to the MEng level. This involves the equivalent of one year's additional full-time study. It may be a specialist MSc focusing on one specific area of engineering. Some employers encourage their trainee engineers to participate in an integrated graduate development scheme (IGDS), a route particularly supported by the Institution of Mechanical Engineers, which includes part-time study at a university to enhance the training they are receiving during their employment.

Graduates in other degree subjects such as physics, chemistry, geology and mathematics can also become chartered engineers by taking additional qualifications such as an appropriate Master's in engineering or participating in an employer's integrated development scheme.

City & Guilds runs the 'Engineering Council Examination' which is in two parts and this also provides a route to meeting the qualification requirements for chartered engineer status.

Professional development

Having completed the academic requirements equivalent to a MEng degree, aspiring chartered engineers must then undertake a period of 'initial professional development' en route to professional status. During this time they build their competence and professional depth and relate the technical aspects of their role to the requirements of the business they are working in. The Engineering Council (UK) expects that they will learn to 'take account of social and economic factors in their decision making'.

Many employers provide structured professional development programmes that are accredited by the engineering institutions and approved by the Engineering Council (UK). These programmes usually include training that is relevant to the environment in which the engineer is working and opportunities to learn 'on the job' by gaining a broad range of experience in an engineering role.

It is expected that trainees will keep a log of all their experiences and eventually present this together with other material at a professional review interview with senior engineers. In the case of civil engineers this is a discussion of the details of a project on which they have been working together with all the plans, calculations and other papers.

Engineers are also expected to spend time in a responsible job that relates strongly to their first key role as a professional. They must demonstrate that they can be accountable for their technical competence before being granted the qualification of chartered engineer.

ADVANTAGES OF PROFESSIONAL BODIES

Joining a professional body has many advantages. They usually produce monthly newsletters or magazines that keep you up to date with developments in your profession. Young engineers are often especially well catered for by the provision of special events, as mentioned earlier in this chapter. Attending conferences or seminars organised by the institutions can give you the opportunity to meet other engineers with similar interests and key people who are at the forefront of your profession. Most institutions have special interest groups that provide a meeting point for professionals working on similar problems in the same field. As a member of an institution you may eventually be able to get involved in organising conferences and seminars for like-minded people, sitting on committees or helping with the editorial side of publications. All of these activities can enhance your professional standing and lead to jobs with enhanced responsibility.

There are plenty of opportunities to get into engineering and, once there, to make significant progress. Initiative and enthusiasm coupled with skills and professionalism are usually rewarded. So if you have a real interest in any part of engineering and the determination to succeed, like Clare Roberts in the Introduction, then you probably will.

USEFUL ADDRESSES

Association of Consulting Engineers
Alliance House
12 Caxton Street
London SW1H 0QL
Tel: 020 7222 0750
www.acenet.co.uk

Association of Licensed Aircraft Engineers Bourn House
8 Park Street
Bagshot
Surrey GU19 5AQ
Tel: 01276 474888
www.alae.mcmail.com

The British Computer Society
1 Sanford Street
Swindon SN1 1HJ
Tel: 01793 417417
www.bcs.org.uk

The British Institute of Non-destructive Testing
1 Spencer Parade
Northampton NN1 5AA
Tel: 01604 231489
www.binstndt.org.uk

The Chartered Institution of Building Services Engineers
222 Balham High Road
London SW12 9BS
Tel: 020 8675 5211
www.cibse.org.uk

The Chartered Institution of Water and Environmental Management
15 John Street
London WC1N 2EB
Tel: 020 7831 3110
www.ciwem.org.uk

EMTA (The Engineering Manufacture Training Organisation)
EMTA House
Watford WD18 0JT
www.emta.org.uk

The Engineering Council (UK)
10 Maltravers Street
London WC2R 3ER
Tel: 020 7240 7891
www.engc.org.uk

The Engineering and Technology Board
10 Maltravers Street
London WC2R 3ER
Tel: 020 7240 7333
www.etechb.co.uk

The Institute of Acoustics
77a St Peters Street
St Albans
Hertfordshire AL1 3BN
Tel: 01727 848195
www.ioa.org.uk

The Institute of Cast Metals Engineers
Bordesley Hall
The Holloway
Alvechurch
Near Birmingham B48 7QA
Tel: 01527 596100
www.icme.org.uk

The Institute of Corrosion
Corrosion House
Vimy Court
Vimy Road
Leighton Buzzard
Bedfordshire LU7 1FG
Tel: 01525 851771
www.icorr.demon.co.uk

The Institute of Energy
18 Devonshire Street
London W1G 7AU
Tel: 020 7580 0077
www.instenergy.org.uk

The Institute of Healthcare Engineering & Estate Management
2 Abingdon Road
Cumberland Business Centre
Northumberland Road
Portsmouth PO5 1DS
Tel: 023 9282 3186
www.iheem.org.uk

The Institute of Highway Incorporated Engineers
20 Queensbury Place
London SW7 2DR
Tel: 020 7823 9093
www.ihie.org.uk

The Institute of Marine Engineering Science & Technology
80 Coleman Street
London EC2R 5BJ
Tel: 020 7382 2600
www.imarest.org.uk

The Institute of Materials
1 Carlton House Terrace
London SW1Y 5DB
Tel: 020 7451 7300
www.materials.org.uk

The Institute of Measurement and Control
87 Gower Street
London WC1E 6AA
Tel: 020 7387 4949
www.instmc.org.uk

The Institute of Physics
76–78 Portland Street
London W1N 4AA
Tel: 020 7470 4800
www.iop.org

The Institute of Physics and Engineering in Medicine
Fairmount House
230 Tadcaster Road
York YO24 1ES
Tel: 01904 610821
www.ipem.org.uk

The Institute of Plumbing
64 Station Lane
Hornchurch
Essex RM12 6NB
Tel: 01708 472791
www.plumbers.org.uk

The Institution of Agricultural Engineers
West End Road
Silsoe
Bedford MK45 4DU
Tel: 01525 861096
www.iagre.org

The Institution of Chemical Engineers
165–189 Railway Terrace
Rugby CV21 3HQ
Tel:01788 578214
www.icheme.org

The Institution of Civil Engineers
1 Great George Street
London SW1P 3AA
Tel:020 7222 7722
www.ice.org.uk

The Institution of Electrical Engineers
Savoy Place
London WC2R 0BL
Tel: 020 7240 1871
www.iee.org.uk

The Institution of Engineering Designers
Courtleigh
Westbury Leigh
Westbury
Wiltshire BA13 3TA
Tel: 01373 822801
www.ied.org.uk

The Institution of Fire Engineers
148 Upper New Walk
Leicester LE1 7QB
Tel: 0116 255 3654
www.ife.org.uk

The Institution of Gas Engineers
21 Portland Place
London W1N 3AF
Tel: 020 7636 6603
www.igase.org.uk

The Institution of Incorporated Engineers
Savoy Hill House
London WC2R 0BS
Tel: 020 7836 9006
www.iie.org.uk

The Institution of Lighting Engineers
Lennox house
9 Lawford Road
Rugby CV21 2DZ
Tel: 01788 576492
www.ile.co.uk

The Institution of Mechanical Engineers
1 Birdcage Walk
London SW1H 9JJ
Tel: 020 7222 7899
www.imeche.org.uk

The Institution of Mining and Metallurgy
Danum House
South Parade
Doncaster DN1 2DY
Tel: 01302 320486
www.imm.org.uk

The Institution of Nuclear Engineers
Allan house
1 Penerley Road
London SE6 2LQ
Tel: 020 8698 1500
www.inuce.org.uk

The Institution of Railway Signal Engineers
Savoy Hill House
Savoy Place
London WC2R 0BS
Tel: 020 7240 3290
www.irse.org.uk

The Institution of Structural Engineers
11 Upper Belgrave Street
London SW1X 8BH
Tel: 020 7235 4535
www.istructe.org.uk

The Institution of Water Officers
4 Carlton Court
Team Valley
Gateshead
Tyne & Wear NE11 0AZ
Tel: 0191 422 0088
www.iwohq.demon.co.uk

The Royal Aeronautical Society
4 Hamilton Place
London W1V 0BQ
Tel: 020 7499 6230
www.raes.org.uk

The Royal Institution of Naval Architects
10 Upper Belgrave Street
London SW1X 8BQ
Tel: 020 7235 4622
www.rina.org.uk

The Society of Environmental Engineers
The Manor House
Buntingford
Hertfordshire SG9 9AB
Tel: 01763 271209
www.environmental.org.uk

The Society of Operations Engineers
22 Greencoat Place
London SW1P 1PR
Tel: 020 7630 1111
www.soe.org.uk

Universities and Colleges Admissions Service (UCAS)
Rosehill
New Barn Lane
Cheltenham
Gloucestershire GL52 3LZ
Tel: 01242 227788
www.ucas.com

The Welding Institute
Granta Park
Great Abington
Cambridge CB1 6AL
Tel: 01223 891162
www.twi.co.uk

The Year in Industry
University of Manchester
Simon Building
Oxford Road
Manchester M13 9PL
www.yini.org.uk